T0114384

FLUIDISED
PARTICLES

TO
THE MEMORY OF
T. R. C. FOX

FLUIDISED PARTICLES

BY

J. F. DAVIDSON

M.A., Ph.D., A.M.I.Chem.E., A.M.I.Mech.E.

Fellow of Trinity College and Lecturer in Chemical Engineering in the University of Cambridge

AND

D. HARRISON

M.A., Ph.D., A.M.I.Chem.E., F.R.I.C.

Fellow of Selwyn College and Lecturer in Chemical Engineering in the University of Cambridge

CAMBRIDGE

AT THE UNIVERSITY PRESS

1963

CAMBRIDGE
UNIVERSITY PRESS

University Printing House, Cambridge CB2 8BS, United Kingdom

One Liberty Plaza, 20th Floor, New York, NY 10006, USA

477 Williamstown Road, Port Melbourne, VIC 3207, Australia

314-321, 3rd Floor, Plot 3, Splendor Forum, Jasola District Centre, New Delhi - 110025, India

79 Anson Road, #06-04/06, Singapore 079906

Cambridge University Press is part of the University of Cambridge.

It furthers the University's mission by disseminating knowledge in the pursuit of education, learning and research at the highest international levels of excellence.

www.cambridge.org
Information on this title: www.cambridge.org/9780521047890

© Cambridge University Press 1963

First published 1963

A catalogue record for this publication is available from the British Library

ISBN 978-0-521-04789-0 Hardback

CONTENTS

PREFACE   *page* ix

SYMBOLS AND NOTATION   xii

1. The Fluidised State

1.1 Introduction   1

1.2 The use of fluidised systems   1

1.3 Definitions   3

1.4 The range of the fluidised state   5

1.5 The incipient fluidising velocity U_0 8

1.6 Expansion of a particulately fluidised bed   15

1.7 The two-phase theory of fluidisation   19

2. The Rise and Coalescence of Bubbles in Fluidised Beds

2.1 Introduction   21

2.2 Bubbles and drops in liquids   21

2.3 The rate of rise of bubbles in fluidised beds   29

2.4 The slugging behaviour of fluidised beds   41

2.5 The coalescence of bubbles in fluidised beds   42

3. The Formation of Bubbles in Fluidised Beds

3.1 The analogy with the gas–liquid system   50

3.2 The theory of bubble formation in an inviscid liquid 50

3.3 Experimental results on bubble formation with a gas–water system   53

3.4 Experimental results on bubble formation at a single orifice in a fluidised bed   57

3.5 Bubble formation in a fluidised bed with distributed gas supply   61

4. The Exchange Between the Bubble and Particulate Phases

4.1 Introduction *page* 63

4.2 The relative motion between fluid and particles in the particulate phase 63

4.3 Motion of the particles, and of the fluid, around a rising bubble 66

4.4 Interpretation; particle and fluid streamlines; comparison with experiment 70

4.5 Pressure distribution and inter-particle forces around a rising bubble 74

5. The Stability of Bubbles in Fluidised Beds

5.1 The transition between aggregative and particulate fluidisation 80

5.2 The stability of fluidised beds 81

5.3 The maximum size of stable bubble in a fluidised bed 84

5.4 The relation of the theory of bubble stability to experiment 87

5.5 Further consideration of the theory of bubble stability 93

6. The Fluidised Bed as a Catalytic Reactor

6.1 Introduction 97

6.2 Catalytic conversion with a first-order reaction 98

6.3 Comparison with experimental results for a first-order reaction 106

6.4 Discussion of the theoretical models 118

6.5 The importance of heat and mass transfer in a fluidised catalytic reactor 121

APPENDIX A

A. 1 Irrotational motion past a cylinder or sphere   *page* 123

A. 2 Percolation of fluid through a void in a particle bed   129

APPENDIX B

The pressure recovery in the wake below a spherical-cap
bubble   132

APPENDIX C

Diffusion from the curved surface of a spherical-cap bubble   134

BIBLIOGRAPHY   139

INDEX OF AUTHORS   145

GENERAL INDEX   147

PREFACE

This book is the outcome of five years' work on the fluidisation of solid particles; although the work is far from complete, the stage has been reached at which a reasonably connected picture can be given of the mechanism of fluidisation. The main thesis advanced is that the behaviour of fluidised beds can be explained in terms of the behaviour of bubbles of fluidising fluid within the bed of particles. This leads to a relatively simple picture of an aggregatively fluidised bed, and to a possible explanation of the difference between aggregative and particulate fluidisation. The treatment brings out the remarkable similarity between the behaviour of bubbles in a fluidised bed and in an ordinary liquid.

The scope of the book is deliberately limited; attention is concentrated on a bed of particles in a fixed vessel and fluidised by gas or liquid, the gas–solids system being of major interest because of its considerable industrial importance. These terms of reference exclude, for example, the pneumatic conveying of particles, and heat transfer from a fluidised bed to the walls of the containing vessel. However, it is believed that these phenomena will come to be explained on the basis of the principles here set forth. The rheology of powders—which some authors regard as a branch of fluidisation—is also not included: at the present time this seems to stand more as a subject by itself. On the other hand, fluidisation might be regarded as a branch of rheology, a fluidised bed of particles being a rheological fluid with special properties.

An attempt has been made to make the book complete in itself by providing, where necessary, analysis of a supporting and ancillary kind in appendices. Nevertheless, it is hoped that the thread of the argument in the main text may be followed without reference to the appendices. Some parts of the theory are published here for the first time, in particular the discussion in Chapter 4 of the pressure distribution in the neighbourhood of a rising bubble, and the analysis of data in the literature on catalytic reactors, given in Chapter 6. Some of the ideas set out are of a speculative nature, and to some degree they are presented as a stimulus to further research. An example of this is the theory given in Chapter 5 concerning the stability of bubbles in fluidised beds, for this has not been established in detail, although it appears to give a reasonable semi-quantitative explanation for the differences between aggregative and particulate fluidisation.

Although the study of fluidisation has not reached the stage at which large reactors can be designed solely from first principles, a number of results are given which should aid the process designer or plant operator. However, it is hoped that the main help such readers will derive from the text is a physical understanding of fluidisation phenomena. Such an understanding is ultimately far more important than a knowledge of a dimensionless correlation derived from laboratory-size experiments; for this type of correlation is not often based on a physical appreciation of the processes involved, and in any case should not be used in conditions which differ from those of the experiments from which it was derived. It has been possible to include only a few data from large reactors; data on large fluidised reactors are as conspicuous by their absence from the literature as the reactors themselves are conspicuous by their presence in oil refineries, but it is felt that the theory presented is sufficiently based on first principles to allow it to be applied with some confidence to full-scale systems.

For those who teach Chemical Engineering, and for students of it, this book should provide examples of the application of the principles of fluid mechanics to problems of genuine chemical engineering interest. Much of today's textbook theory of fluid mechanics was inspired by the study of hydraulics and aeronautics; much of the theory given in the following pages was inspired by the study of processes of importance to chemical engineers, and therefore it is hoped that some of it—for example, that on bubble rise and bubble formation at an orifice—will find a central place in courses on fluid mechanics for chemical engineers. Although only part of the book is suitable for undergraduate teaching, the whole of it might form the background to a postgraduate course, where its incomplete nature should provide the right challenge for research.

We would like to acknowledge our debt to the research students who have done so much of the work we describe; their energy and enthusiasm have been a constant source of inspiration. We are grateful to Dr K-A. Melkersson for helpful criticism of the draft at an early stage, and to Dr N. Epstein for suggestions with regard to Chapter 1. We would also like to thank Miss M. Sansom for her skilful typing of the manuscript. One of us (J.F.D.) would like to acknowledge the support of the University of Delaware during a visit in 1960 when the ideas for much of Chapter 6 were formulated. We are indebted to Dr P. N. Rowe for providing Plates II, III and V, which are published by kind permission of the United Kingdom Atomic Energy Authority, and to Mr J. Hardebol for providing Plate XI, which is published by kind permission of the Institution of Chemical Engineers.

J. F. D.

CAMBRIDGE

D. H.

August 1962

SYMBOLS AND NOTATION

A	radius of circle of penetration; cross-sectional area of tube
A_m	cross-sectional area of passage
a	radius of bubble; surface area per unit volume of packing
B	arbitrary constant in Bernoulli's theorem
b	radius of cylinder or sphere
$C_1 C_2$	arbitrary constants
C_i	inlet concentration of carbon tetrachloride (Szekely, 1962)
C_0	concentration of cumene (Mathis and Watson, 1956); outlet concentration of carbon tetrachloride (Szekely, 1962)
c	concentration at any point; constant defined by (2.23)
c'	c_H/c_0
c^*	surface concentration of solute gas
c_0	concentration in bubble and at bed entry
c_b	concentration within bubble
c_{bH}	concentration within bubbles leaving the top of the bed
c_D	drag coefficient for an isolated sphere
c'_D	drag coefficient for a particle in a packed bed
c_f	pipe friction coefficient
c_H	concentration of mixed gas leaving bed
c_p	concentration within particulate phase
c_{pH}	concentration within the gas leaving the particulate phase
D	tube diameter
D_e	diameter of sphere having the bubble volume
D_{em}	maximum value of D_e
D_f, D_{f1}, D_{f2}	frontal diameter of bubble
D_G	gas-phase diffusion coefficient
d	particle diameter
d_H	hydraulic mean diameter = $4 \times$ area/perimeter
F	force on an isolated sphere; flow quantity (Lewis, Gilliland and Glass, 1959)
F'	force on particle in packed bed
Fr	Froude number = U_0^2/gd
f	packing friction factor = $\dfrac{\Delta p d \epsilon^3}{L \rho U^2 (1 - \epsilon)}$
G	gas volume flow-rate

g	acceleration of gravity
H	bed height or height of liquid column
H_0	bed height at incipient fluidisation; height of liquid column with no air flow
H_S	settled height of bed
h	bubble height
J	pressure gradient at infinity
K	permeability constant
K_{AB}	transfer coefficient (Mathis and Watson, 1956)
K_d	transfer coefficient (Massimilla and Johnstone, 1961)
k	reaction velocity constant
k'	kH_0/U
k_d	transfer coefficient (Pansing, 1956)
k_G	mass transfer coefficient between a bubble and its surface
L	length of fixed bed
L_0, L_1, L_2	bubble entrance effects
ΔL	initial distance between bubbles
l	perimeter of passage
M	mass equivalent to the accelerated fluid
m	$U_b/U_{b\infty}$
$\mathrm{d}m$	mass of fluid element
m_1, m_2	roots of quadratic equation
N	number of bubbles per unit volume; number of spheres per unit packed volume
N_c	transfer of diffusing substance from curved surface of bubble
n	frequency of bubbles per sec; index given by (1.20)
P	$p_f + p_p$
ΔP	defined in fig. 29
p	pressure in percolating fluid, and in fluid flowing round spherical-cap bubble
p_0	pressure within bubble
p_f	pressure within fluidising fluid
p_R	pressure recovery below spherical-cap bubble
Δp_f	defined in fig. 29
Δp	pressure difference across length L of packing; partial pressure difference (Pansing, 1956)
p_p	pressure equivalent of inter-particle forces

Q	$q + k_G S$
q	rate of exchange between a bubble and the particulate phase
R	radius of spherical cap; universal gas constant
Re	Reynolds number $= \rho U d / \mu$ or $\rho u_m d_H / \mu$
Re$'$	$\mathrm{Re}/(1 - \epsilon)$
Re$_0$	$\rho U_0 d / \mu$
r	bubble radius; polar coordinate
S	surface area of rising bubble
s	distance travelled from orifice by bubble centre
T	temperature of bed
t	time
Δt	time interval between bubble injections
t_c	coalescence time
t_s	time for bubble to reach surface
U	superficial velocity of fluidising fluid
U_0	superficial velocity of fluidising fluid at incipient fluidisation
U_1, U_2	velocities of leading and following bubbles
U_A	absolute rising velocity of bubble
U_b	rising velocity of bubble in stagnant liquid
$U_{b\infty}$	value of U_b in a large mass of liquid
U_c	upward velocity within bubble
U_R	relative velocity between bubbles
U_t	terminal velocity of particle
U_W	absolute velocity of the wake behind a bubble
u	absolute velocity of fluidising fluid
u_0	interstitial velocity of fluidising fluid at incipient fluidisation
u_m	mean velocity of fluid in passage
u_x, u_y, u_r, u_θ	components of u
V	bubble volume
V_m	maximum volume of bubble
v	particle velocity
v_x, v_y, v_r, v_θ	components of v
W	velocity of ideal fluid
W_s	flow-rate into all bubbles from particulate phase (May, 1959)

w	velocity of ideal fluid
w_x, w_y, w_r, w_θ	components of w
w_s	surface velocity
X	$QH/U_A V$ = number of transfer units
x	horizontal coordinate; fraction of cumene converted (Mathis and Watson, 1956)
y	vertical coordinate; distance from bubble surface
z	vertical coordinate measured from bubble top
α	semi-angle measured round spherical-cap bubble
α_1	maximum value of α
β	$1 - U_0/U$
ϵ	mean voidage fraction of fixed or fluidised bed
ϵ_0	value of ϵ at incipient fluidisation
η	defined in (C.8)
θ	polar coordinate
λ	inclination of passage
μ	viscosity of fluid
ν	kinematic viscosity of fluidising fluid, or of the fluidised bed
ρ	density of fluid
$\Delta\rho$	$\rho_s - \rho_f$
ρ_c	density of continuous phase
ρ_f	density of bubble phase
ρ_L	liquid density
ρ_p	bulk density of particulate phase
ρ_s	density of solid particle
τ_0	wall shear stress
Φ	defined by (C.6)
Φ_1	maximum value of Φ
ϕ	velocity potential
ψ	stream function
ψ_f	stream function of fluidising fluid

CHAPTER I

THE FLUIDISED STATE

1.1. Introduction

The phenomenon of fluidisation can best be visualised in terms of a simple experiment in which a bed of solid particles is supported on a horizontal gauze in a vertical tube. Gas or liquid is then forced to flow upwards through the gauze, and so through the particle bed. This flow causes a pressure drop across the bed, and when this pressure drop is sufficient to support the weight of the particles the bed is said to be *incipiently fluidised*. Any further increase in flow causes the bed to expand to accommodate the increase. The fluidised bed thus formed has many of the properties of a liquid; its upper surface remains horizontal when the containing apparatus is tilted, and it hardly impedes the movement of objects floated on the surface. When the flow of gas or liquid through the bed is increased still further, to the point at which the flow velocity becomes greater than the free-falling velocity of the particles, then, clearly, the particles are carried out of the apparatus.

1.2. The uses of fluidised systems

A fluidised system has a number of highly useful properties, the more important being concerned with temperature control and heat transfer, continuity of operation, and catalytic reactions.

Temperature control and heat transfer

The same temperature is quickly established throughout a fluidised system because the general agitation of the particles disperses local regions of hot or cold. A fluidised bed is therefore very suitable for catalytic reactions requiring close temperature control. There is also a high rate of heat transfer to a solid object placed in the bed, so that it is a very convenient heat transfer medium; for this reason a gas-fluidised bed can be used as a constant temperature bath in which to immerse a reaction vessel that has to be at a high temperature (Adams, Gernand and Kimberlin, 1954).

Continuity of operation

A fluidised system enables solid particles to be handled essentially as a liquid, and this can be a very considerable asset in the design of a continuous process. The addition and withdrawal of solid particles from the process equipment is also facilitated.

Catalytic reactions

Fluidisation is an excellent way of bringing a gas into contact with a solid, and therefore catalytic reactions are often well suited to the technique. Fluidised beds first became of major importance through the development of a fluidised process for cracking heavy hydrocarbons into petroleum spirit (Murphree, Brown, Fischer, Gohr and Sweeney, 1943).

The applications of fluidisation fall into two broad categories: (i) chemical reactions and catalysis, and (ii) physical and mechanical processes. The drying of solid particles (Jobes, 1954) is an example of (ii). Apart from fluid catalytic cracking, the chemical uses of fluidisation have also included the Fischer–Tropsch process (Hall and Crumley, 1952), the roasting of pyrites (Thompson and MacAskill, 1955), and the reduction and fluorination of uranium (Hawthorn, Shortis and Lloyd, 1960). This brief list of applications is only intended to be illustrative; it is by no means exhaustive. Zenz and Othmer (1960) describe some 20 uses of fluidisation, and they refer to 50 more.

The disadvantages of fluidised systems

A fluidised bed is not suitable for all fluid–solids processes, and considerable difficulties have arisen in the past when some of its disadvantages have not been clearly recognised, for instance:

(i) The quick equilibration of temperature in a fluidised system means that it is unsuitable for a reaction which is best carried out in a reactor giving a temperature gradient along the reaction path.

(ii) The ease with which particles can be fluidised can vary enormously, and thus a fluidised process is usually precluded for particles which do not flow freely or which agglomerate (e.g. waxes).

(iii) The bubbles of gas characteristic of most gas–solids beds can cause both chemical and mechanical difficulties. For instance,

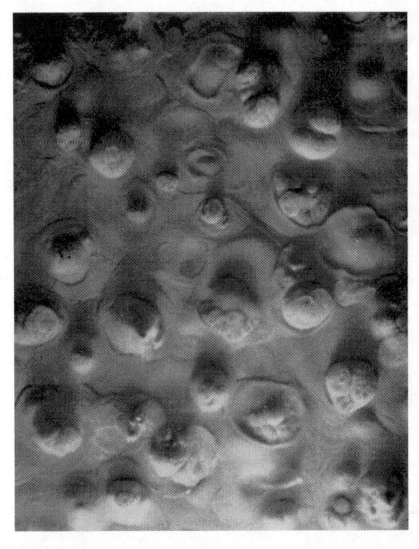

Plate I. Aggregative fluidisation showing bubbles breaking surface at the top of a bed of sand fluidised by air.

(*Facing p.* 3)

in a gas–solids reaction, it is possible for some of the gas in the bubbles to by-pass the particles altogether, and so the overall efficiency of contact is lowered. The bubbles can also cause mechanical buffeting sufficient to cause breakage when, for example, weak fabric or paper is dried in a hot fluidised bed.

1.3. Definitions

(a) Incipient fluidisation

Typical results for a gas-fluidised bed of particles are shown in fig. 1; the pressure drop across the bed, and its height, are plotted as functions of the superficial gas velocity. There is marked hysteresis, so that for slowly increasing flow the curves A are generated, while the curves D are for slowly decreasing flow; the amount of hysteresis depends upon the degree of consolidation of the original bed. At point B, the overall pressure drop is slightly more than enough to support the weight of the particles, owing to the wedging action within the bed. A slight increase in flow above point B frees the particles, the pressure drop becomes just enough to support their weight, and consequently point C is usually defined as the point of *incipient fluidisation*, the superficial gas velocity being U_0 and the voidage fraction, ϵ_0. If the flow rate is now slowly decreased from point C, the particles are more loosely packed, and consequently the bed height is greater and the pressure drop smaller, as shown by curves D. By starting with a bed of loosely packed particles (one that has just been fluidised) it is possible to get the curves for increasing flow-rate to coincide with the curves for decreasing flow-rate; the hysteresis is then eliminated.

Below incipient fluidisation, results similar to those in fig. 1 are obtained for liquid-fluidised beds.

(b) Aggregative fluidisation

For a gas-fluidised bed in which the gas velocity is greater than U_0, some of the gas may pass through the bed as bubbles, and these can be seen to burst when they reach the top surface of the bed as shown in Plate I. This is commonly known as *aggregative fluidisation*, and usually occurs when solids are fluidised by gases. The bubbles agitate the bed, and consequently its height fluctuates as indicated in region E of fig. 1.

(c) Particulate fluidisation

For a liquid-fluidised bed in which the velocity is greater than U_0, the bed height increases with velocity, but there are usually no marked fluctuations in level, the particles spacing themselves evenly so that the liquid passes smoothly through the interstices without the formation of bubbles. This is known as *particulate fluidisation*.

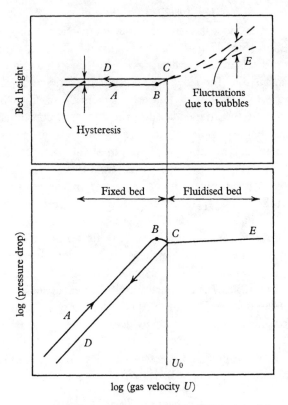

Fig. 1. Typical curves for a gas-fluidised bed of particles of approximately uniform size.

The difference in behaviour between gas- and liquid-fluidised beds may be readily observed by fluidising glass beads of about 0·25 mm diameter in turn with air and water; experimental data on both types of fluidisation are given by Wilhelm and Kwauk (1948). Although aggregative fluidisation is normally found with gas–

solids systems and particulate behaviour with liquid–solids systems, there are exceptions to this generalisation. For example, Wilhelm and Kwauk (1948) found that a bed of lead shot fluidised by water behaved aggregatively (i.e. water bubbles were present); and Leung (1961) has observed particulate behaviour when fluidising light resin particles with carbon dioxide under pressure. The detailed relationship between aggregative and particulate fluidisation is considered in Chapter 5.

1.4. The range of the fluidised state

The general appearance of an aggregative bed as the flow of fluid is increased is shown diagrammatically in fig. 2. First, the bed

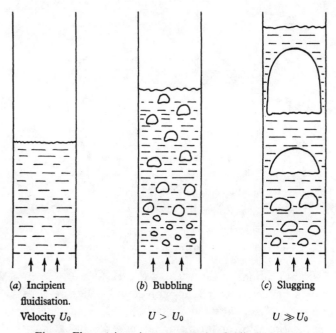

(a) Incipient fluidisation. Velocity U_0

(b) Bubbling $U > U_0$

(c) Slugging $U \gg U_0$

Fig. 2. Flow régimes in an aggregative fluidised system.

expands from a fixed bed to the point of incipient fluidisation and then, as the flow-rate is further increased, bubbling begins. With still greater flows the bubbles grow and appear more frequently, until their frontal diameters are equal to the diameter of the containing apparatus. The bed, shown in fig. 2(c), is then said to be *slugging*. A further increase of flow to the bed carries the particles out of the apparatus. The voidage fraction is then high (over 0·80), and the phenomenon is that of pneumatic transport, or *dilute-phase fluidisation*.

The general appearance of particulate systems is shown in fig. 3. In this case the bed is always homogeneous, with uniform expansion to take up the increased flow, dilute-phase fluidisation being reached without the formation of bubbles. The expansion of a particulate bed is considered in detail in § 1.6, p. 15.

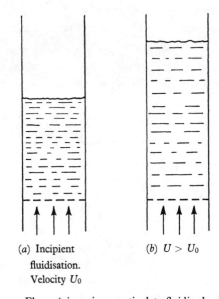

(a) Incipient
fluidisation.
Velocity U_0

(b) $U > U_0$

Fig. 3. Flow régimes in a particulate fluidised system.

Most applications of fluidisation at present are concerned with gas–solids systems rather than with liquid–solids systems; and a quantitative consideration of the bubbling and slugging regions of gas-fluidisation takes up the major part of this book. However, there are other important aspects of gas-fluidised behaviour which still await analytical attention, for example:

(a) Channelling and spouted beds

Channelling is the phenomenon observed when a disproportionately large amount of the fluidising fluid follows one or two particular paths through the bed. This is often a marked characteristic of a bed of very fine particles, or of sticky or waxy particles which tend to agglomerate.

A spouted bed may be seen as an extreme form of channelling, in which the fluidising fluid takes only one (usually central) path up the bed. An account of spouted beds has been given by Leva (1959).

(b) Distributor design

Little systematic work has been reported on the influence of the bed support on fluidised behaviour; although many data have been collected by industry on the design of distributors for specific purposes. Grohse (1955) discovered that the point of incipient fluidisation was more reproducible with a porous plate distributor than if the bed was supported on either a 300 BS mesh ($\simeq$ 0·005 cm spacing) screen or a multi-orifice plate. Rowe and Stapleton (1961) observed the behaviour of a gas-fluidised bed of 12 in diameter fitted in turn with a 'bubble-cap' distributor, a conical distributor, and a porous plate. They too found that the porous plate distributor allowed a more even expansion of the bed than the other distributors, and that it gave rise to more—and smaller—bubbles. They also found that the distributor design affected the behaviour of the bed over most of its height.

(c) Baffles

The use of 'baffles' to promote even fluidisation is another subject on which published data are sparse. It was discovered early in the study of fluidised beds that internal baffles tend to break up bubbles and, as a result, provide smoother operation. A strikingly successful example of this was the smooth fluidisation of a bed 80 in deep and 1 in diameter by Hall and Crumley (1952). The baffle arrangement consisted of discs of 10 BS mesh ($\simeq$ 0·15 cm) steel gauze, dished slightly convex to the gas flow, attached at 2 in intervals to a vertical rod in the centre of the bed. This type of baffle impeded bubble coalescence over the full depth of the bed, and thus the slugging of the bed, which occurred without the baffles, was avoided.

Beck (1949), Massimilla and Westwater (1960), and Volk, Johnson and Stotler (1962), have also investigated baffled beds and, aside from the effect on bubbling, two other general points emerge from their work:

(i) The average linear velocity of the particles in the bed is reduced, in some cases to 10–20 % of the velocity in an unbaffled bed; and

(ii) heat transfer between the bed and the containing walls is adversely affected, possibly as a consequence of (i).

At the present time the precise design of baffles for a particular system allows considerable scope for trial-and-error ingenuity, and so the patent literature on this subject is more extensive than the academic (Hassett, 1963).

(d) The fluidisation of very large particles

Squires (1962) distinguishes between beds with particles of a size smaller than about 20 BS mesh ($\simeq$ 0·08 cm) ('fluid beds') from those with particles larger than about 10 BS mesh ($\simeq$ 0·15 cm) ('Teeter beds'). The work described in this book is concerned in the main with gas- and liquid-fluidised systems containing smaller particles of less than 10 BS mesh.

1.5. The incipient fluidising velocity U_0

There is no doubt that the best way to determine U_0 is to measure it. The method is to measure the pressure drop through the bed of particles as a function of the flow-rate for slowly increasing and then slowly decreasing flow-rates. The results give curves of the kind shown in fig. 1, and U_0 is the velocity at point C, though this point is not always well defined. An experiment of this kind can be done on the laboratory scale and the results can be applied with reasonable confidence to a large-scale plant, provided the pressure and temperature are the same.

Nevertheless, it is useful to be able to estimate U_0 from first principles, both for design purposes, and because the plant conditions may be difficult to simulate in the laboratory if they involve high pressure or temperature. Since the individual particles derive mutual support from one another for $U < U_0$, the problem of predicting U_0 is essentially the problem of finding the flow which will produce a pressure drop through the fixed particle bed equal to its weight per unit cross-section. For this purpose a short account will be given of the theory of flow through fixed beds of particles.

(a) Pressure drop through a fixed bed of particles

In this section we shall consider the flow of fluid through a fixed bed of particles under the influence of a uniform pressure gradient.

The following theory was originated by Kozeny (1927) and Carman (1937), and to take the simplest view of it, the idea is to obtain an equivalence between the tortuous passages through the packing and a single passage having the same volume and surface

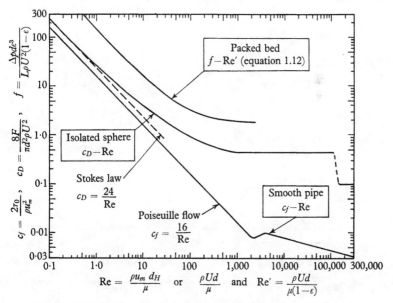

Fig. 4. Relations between pressure drop and flow for a smooth pipe, an isolated sphere, and a packed bed.

area. When fluid flows through a parallel-sided passage, the friction coefficient, c_f, is uniquely related to the Reynolds number Re, where

$$c_f = 2\tau_0/\rho u_m^2, \tag{1.1}$$

and

$$\text{Re} = \rho u_m d_H/\mu, \tag{1.2}$$

τ_0 being the wall shear stress, ρ and μ the fluid density and viscosity, u_m the mean velocity in the passage and d_H its hydraulic mean diameter. Fig. 4 shows the relation between c_f and Re for a circular pipe (e.g. Prandtl, 1952, p. 165) and a similar relation between corresponding quantities f and Re′ for a packed bed of spheres. The friction factor f is derived from c_f, and Re′ from Re by finding d_H in terms of the packing geometry, and u_m in terms of U, the superficial velocity through the packing.

Fig. 5 shows the passage equivalent to the spaces within the particles contained in a volume having length L in the direction of flow, and unit cross-section normal to this direction. The passage is inclined at angle λ to the direction of the mean flow, and its cross-sectional area is A_m, and the perimeter of that cross-section is l. The

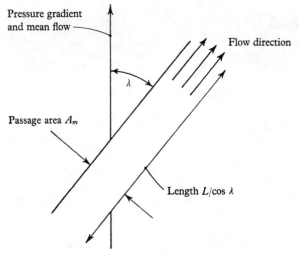

Fig. 5. Inclined passage having the same flow resistance as a packed bed.

volume of the passage, $LA_m/\cos \lambda$, is equal to the void volume $L\epsilon$ within the packed volume, so that

$$\epsilon = A_m/\cos \lambda. \tag{1.3}$$

The hydraulic mean diameter of the passage is $4A_m/l$; now the surface area of the passage wall, $Ll/\cos \lambda$ is assumed equal to the surface area of the particles La, a being the surface area per unit packed volume. Therefore, using (1.3), the hydraulic mean diameter of the passage is

$$d_H = \frac{4A_m}{l} = \frac{4\epsilon}{a}, \tag{1.4}$$

and this is taken as the hydraulic mean diameter of the spaces between the packing.

The flow within the passage shown in fig. 5 is $u_m A_m = U$, where U is the superficial velocity through the packing and therefore the volume flow-rate through unit cross-section of packing; using (1.3), this equation becomes

$$u_m = \frac{U}{\epsilon \cos \lambda}. \tag{1.5}$$

The flow is caused by the pressure gradient $\Delta p/L$ acting in the direction shown in fig. 5, Δp being the pressure difference across a length L of packing. The pressure gradient causes a mean shear stress τ_0 to act on the passage walls, and since their area is $Ll/\cos \lambda$,

$$\tau_0 = \frac{\Delta p A_m \cos \lambda}{Ll} = \frac{\Delta p \epsilon \cos \lambda}{La}, \tag{1.6}$$

using (1.4). Substituting from (1.5) and (1.6) into (1.1) then gives the friction coefficient for the packing,

$$c_f = \frac{2\Delta p}{\rho U^2 L}\left(\frac{\epsilon^3 \cos^3 \lambda}{a}\right). \tag{1.7}$$

Similarly, substituting from (1.4) and (1.5) into (1.2) gives the Reynolds number for the packing,

$$\text{Re} = \frac{\rho U}{\mu}\left(\frac{4}{a \cos \lambda}\right). \tag{1.8}$$

For low Reynolds numbers, the Poiseuille equation for streamline flow in a pipe is $c_f = 16/\text{Re}$ (Prandtl, 1952, p. 98), and therefore substituting from (1.7) and (1.8) we should expect that

$$\frac{\Delta p}{L} = \frac{2\mu U a^2}{\epsilon^3 \cos^2 \lambda}. \tag{1.9}$$

In particular, for a bed of spheres containing N spheres, of diameter d, per unit volume, $a = \pi d^2 N$ and $1 - \epsilon = \frac{1}{6}N\pi d^3$, giving

$$a = 6(1 - \epsilon)/d. \tag{1.10}$$

Hence from (1.9),

$$\frac{\Delta p}{L} = \left[\frac{72}{\cos^2 \lambda}\right]\frac{\mu U(1 - \epsilon)^2}{d^2 \epsilon^3}. \tag{1.11}$$

For spheres, Carman (1956, p. 14) gives values between 162 and 184 for the square-bracketed term in (1.11), so that λ is between 48° and 51°, which are reasonable values for the average inclination between the fluid streamlines and the direction of mean flow.

For higher Reynolds numbers various workers have used groups like (1.7) and (1.8), with the object of embracing the behaviour of a wide range of packings by means of a single correlation. The first of these was Blake (1922), who derived the groups $\Delta p \epsilon^3 / L \rho U^2 a$ and $\rho U / \mu a$ by dimensional analysis. Ergun (1952) used essentially the same groups $f = \Delta p d \epsilon^3 / L \rho U^2 (1 - \epsilon)$ and $\text{Re}' = \rho U d / \mu (1 - \epsilon)$, where d is given by (1.10), and represents an equivalent sphere having the same surface area per unit volume as the original packing. Fig. 4 shows Ergun's correlation between f and Re' which is

$$f = \frac{150}{\text{Re}'} + 1 \cdot 75. \qquad (1.12)$$

The first term predominates at low Reynolds numbers, and is essentially the same as Carman's form of (1.11), but with a smaller numerical constant. At high Reynolds numbers (up to 2000) the second term in (1.12) predominates, and represents the form drag on the individual particles. The two terms in (1.12) thus represent the effects of viscous and inertia forces, respectively.

In conclusion it must be stressed that this section provides only a brief discussion of fluid flow through fixed beds of particles, and for further information and references to the very voluminous literature on the subject, the reader should consult the works of Brown (1950), Scheidegger (1957), Carman (1956) and Collins (1961). The relationship between this work and pressure drop in fluidised beds has also been considered by Andersson (1961).

(b) An isolated sphere in a uniform stream

This section considers the drag on a sphere in a uniform stream; this drag is of importance in fluidisation in determining the velocity at which the individual particles are carried out of the bed by the fluidising fluid.

Fig. 4 shows a semi-empirical relationship (Coulson and Richardson, 1955, p. 486) between the drag coefficient $c_D = 8F / \pi d^2 \rho U^2$ and Reynolds number $\text{Re} = \rho U d / \mu$, where F is the force on the sphere and U is the velocity of the stream. In the region of very low Reynolds number, Stokes law holds good (Prandtl, 1952, p. 106), so that $c_D = 24/\text{Re}$, and the drag force is

$$F = 3\pi\mu U d. \qquad (1.13)$$

At high Reynolds numbers (10^3 to 10^5) c_D is sensibly constant, though a sharp drop in c_D occurs for Re just greater than 10^5, but this drop is unimportant in the present context.

The similarity between the curves in fig. 4 for an isolated sphere and for a packed bed suggests that the latter is more closely related to the former than to the flow through a pipe as was assumed in the derivation in §1.5 (a), p.9. In particular, the deviation from the straight-line relationship for viscous flow occurs gradually between Re $\simeq$ 1 and 1000 for the single sphere and for the packed bed; the corresponding breakdown of streamline flow in a pipe occurs suddenly at Re $\simeq$ 2000. The gradual transition from purely viscous flow for a sphere is due to the breakaway of the flow pattern behind the sphere; similar breakaway presumably occurs behind each element in a packed bed. This causes the form drag to become more important than the drag due to skin friction, the form drag predominating when the drag coefficient becomes constant at high Reynolds numbers.

(c) Theoretical prediction of U_0

The main difficulty in predicting U_0 is to know the value of the voidage fraction, ϵ_0, at incipient fluidisation. It seems reasonable to suppose that the particles set themselves in the loosest possible mode of packing, whilst yet remaining in contact with one another; if this were true we would expect a bed of uniform spheres to take up the cubic mode of packing, with $\epsilon_0 = \frac{1}{6}(6-\pi) = 0.476$, at incipient fluidisation. This is approximately correct, since voidages of between 0.4 and 0.5 at incipient fluidisation have been observed for spherical particles of uniform size (Leva, 1959, p. 21; Zenz, 1957a).

By assuming that $\epsilon_0 = 0.476$, and substituting it in (1.11) with Carman's average value of $72/\cos^2 \lambda = 180$, we get

$$\frac{\Delta p}{L} = \frac{459\mu U}{d^2}. \tag{1.14}$$

At incipient fluidisation the pressure drop is enough to support the weight of the particles in unit cross-section, and therefore

$$\frac{\Delta p}{L} = (\rho_s - \rho)g(1 - \epsilon_0), \tag{1.15}$$

with ρ_s the density of the material forming the particles. Combining (1.14) and (1.15), with $\epsilon_0 = 0\cdot476$, gives

$$U_0 = 0\cdot00114 g d^2 (\rho_s - \rho)/\mu. \tag{1.16}$$

A similar procedure was followed by Leva (1959, p. 64) who used experimental values of the voidage at incipient fluidisation, and his expression is

$$U_0 = [0\cdot0007\,\mathrm{Re}_0^{-0\cdot063}] g d^2 (\rho_s - \rho)/\mu, \tag{1.17}$$

where $\mathrm{Re}_0 = \rho U_0 d/\mu$. The square-bracketed term was derived empirically from data for incipient fluidisation; this term is of the same order as the coefficient $0\cdot00114$ in (1.16) because Re_0 ranges from 10^{-2} to 10^2 for most fluidised systems, and therefore $\mathrm{Re}_0^{-0\cdot063}$ is of order unity. The difference between the coefficients in (1.16) and (1.17) is probably due to the fact that ϵ_0 is somewhat less than the value of $0\cdot476$ used in deriving (1.16). Alternatively, it may be that there is an inherent difference between a fixed and a fluidised bed which gives the latter a slightly lower resistance; this is discussed in § 1.6 (b), p. 16.

An alternative method, which yields a result very similar to (1.17), has recently been published by Rowe (1961). He used the results of experiments in which water flowed through a regular array of spheres, and the drag force on one sphere was measured by a weighing technique. It was found that the force on the single sphere in the array was $68\cdot5$ times the force on an isolated sphere at the same superficial velocity. Rowe then assumed that this same factor of $68\cdot5$ applies at incipient fluidisation when the drag force just balances the net downward force on the particle. When the Reynolds number is low, (1.13) applies, and therefore

$$68\cdot5 \times 3\pi\mu U_0 d = (\rho_s - \rho)\,\pi d^3 g/6,$$

which simplifies to

$$U_0 = 0\cdot00081(\rho_s - \rho) g d^2/\mu,$$

which is essentially the same as (1.17) since the latter was based on experimental results with a scatter of at least 10 %.

At higher Reynolds numbers, when the Stokes law no longer applies, Rowe found that the factor of $68\cdot5$ still gives a reasonably accurate prediction of the minimum fluidisation velocity. That this is to be expected can be demonstrated by comparing the drag coefficients for a packed bed and for an isolated sphere. For a packed bed with N spheres per unit volume, the drag force per sphere is $F' = \Delta p/NL$ and $N = 6(1-\epsilon)/\pi d^3$, so that $F' = \Delta p \pi d^3/6L(1-\epsilon)$. Now defining a drag coefficient for the individual spheres in the packed bed, $c_D' = 8F'/\pi d^2 \rho U^2$, the friction factor is

$$f = \frac{\Delta p d \epsilon^3}{L\rho U^2(1-\epsilon)} = \tfrac{3}{4}\epsilon^3 c_D'. \qquad (1.18)$$

From (1.12), at high Reynolds numbers (> 1000), the friction factor f for a packed bed is 1.75, and therefore $c_D' = 4 \times 1.75/3\epsilon^3$; at the same Reynolds numbers, the drag coefficient for an isolated sphere $c_D = 0.44$ (fig. 4), and therefore the ratio

$$c_D'/c_D = (4 \times 1.75)/(3 \times 0.44\epsilon^3).$$

If this expression is equated to Rowe's figure of 68.5, $\epsilon_0 = 0.43$, which is in good agreement with the accepted range of values for ϵ_0 (0.4–0.5) for incipiently fluidised spheres of uniform size. It may be noted that the value of ϵ_0 is not very sensitive to changes in c_D'/c_D. In fact any value of this ratio between 40 and 80 will predict ϵ_0 lying within the acceptable range from 0.4 to 0.5.

Thus it appears that Rowe's figure of 68.5, for the ratio of the drag at incipient fluidisation to the drag for an isolated particle, gives a good way of predicting incipient fluidisation. However, it is very surprising that this value was obtained from experiments on the close-packing mode for spheres, when the voidage is only 0.26; moreover, experiments by Martin, McCabe and Monrad (1951), on the pressure drop through regularly packed spheres of that voidage, give a much larger force per sphere than Rowe's value. The drag experiments of Rowe and Henwood (1961) and the pressure-drop measurements of Martin et al. (1951) on more open packings of spheres appear more truly representative of conditions at incipient fluidisation. Values of c_D'/c_D are then less than 68.5, but still large enough to give reasonable figures for ϵ_0.

1.6. Expansion of a particulately fluidised bed

When the superficial velocity U of the fluidising fluid is greater than the incipient velocity U_0, the bed expands uniformly to take up the increased flow, the particles spacing themselves out so that the drag on each is equal to the net weight allowing for buoyancy. Various authors, for example, Wilhelm and Kwauk (1948), and Lewis, Gilliland and Bauer (1949), have measured the expansion of the bed as a function of U. The most satisfactory way of presenting such data is to correlate U/U_t and the mean voidage ϵ, because when $\epsilon \to 1$, $U/U_t \to 1$, since at $\epsilon = 1$ we have a single particle in an infinite fluid. Wilhelm and Kwauk (1948) and Happel and Epstein (1954) gave a graphical relation between U/U_t and ϵ; Richardson and Zaki (1954) gave the result

$$U/U_t = \epsilon^n, \tag{1.19}$$

on the basis of fluidisation and sedimentation experiments. For large tubes, the index n is given by

$$\left. \begin{aligned} n &= 4\cdot65, & \mathrm{Re} &< 0\cdot2, \\ n &= 4\cdot35\,\mathrm{Re}^{-0\cdot03}, & 0\cdot2 &< \mathrm{Re} < 1, \\ n &= 4\cdot45\,\mathrm{Re}^{-0\cdot1}, & 1 &< \mathrm{Re} < 500, \\ n &= 2\cdot39, & \mathrm{Re} &> 500. \end{aligned} \right\} \tag{1.20}$$

Richardson and Zaki found that n is also a function of the ratio of particle to tube diameter, but here we shall consider only large tubes, for which (1.20) holds good.

(a) $\epsilon \to 1$

Although (1.19) does have the right properties when $\epsilon = 1$, in that U is then equal to the free-falling velocity U_t of an isolated particle, Adler and Happel (1962) and Happel (1958) have shown that in the very dilute region, when ϵ is nearly 1, U/U_t should be of the form 1—constant $\times (1-\epsilon)^{\frac{1}{3}}$. This means that the particles have a very marked influence on one another even when they are far apart. Such behaviour does not show up in the results of Richardson and Zaki, because in their experiments ϵ was always less than about $0\cdot95$. However, a figure of $0\cdot95$ is likely to be the upper limit for ϵ in the practical applications of fluidisation.

(b) Comparison with fixed-bed data

If a particulately fluidised bed behaved in the same way as a fixed bed of the same voidage, it would be possible to predict the relation between U and ϵ, that is, the result of Richardson and Zaki, from the correlations for fixed beds, provided these correlations were valid at all values of ϵ. As will be shown, the fixed-bed correlations yield a result similar to that of Richardson and Zaki, though of a rather different form.

We compare an isolated particle, supported by a velocity U_t, with a bed of particles supported by a velocity U; in each case the drag on the individual particle is the same, and the ratio of the drag coefficients is

$$\frac{c_D'}{c_D} = \frac{U_t^2}{U^2}. \tag{1.21}$$

(i) For small Reynolds numbers, Stokes law applies, and $c_D = 24/\mathrm{Re} = 24\mu/\rho U_t d$. For the packed bed at the small Reynolds numbers, (1.12) and (1.18) give

$$f = \tfrac{3}{4}\epsilon^3 c_D' = 150/\mathrm{Re}' = 150\mu(1-\epsilon)/\rho U d.$$

Using this with (1.21) gives

$$\frac{U}{U_t} = \frac{3\epsilon^3}{25(1-\epsilon)}. \tag{1.22}$$

Table 1 shows a comparison between (1.22) and the corresponding expression of Richardson and Zaki, $\epsilon^{4\cdot65}$ (1.20), using the value of n for $\mathrm{Re} < 0\cdot2$. A comparison rather similar to that shown in table 1 was made by Happel (1958).

(ii) At the high Reynolds numbers, c_D tends to $0\cdot44$ (fig. 4), and for a packed bed $f(=\tfrac{3}{4}\epsilon^3 c_D')$ tends to $1\cdot75$ (1.12), so that using (1.21),

$$\frac{U}{U_t} = \left(\frac{3\times0\cdot44}{4\times1\cdot75}\right)^{\frac{1}{2}} \epsilon^{\frac{3}{2}} = 0\cdot434\epsilon^{\frac{3}{2}}. \tag{1.23}$$

Table 1 shows a comparison between (1.23) and the corresponding expression of Richardson and Zaki, $\epsilon^{2\cdot39}$ (1.20), for large Reynolds numbers.

Table 1 shows that there is reasonable agreement between the results of Richardson and Zaki for particulately fluidised beds and the corresponding results (1.22) and (1.23) derived from fixed-bed data. However, table 1 does show the following systematic differences:

(i) Clearly the fixed-bed correlation does not work when $\epsilon \to 1$, for then $U/U_t \to \infty$ for the small Reynolds numbers. But the fixed-bed correlations are based on data for voidages less than about $0\cdot8$;

moreover Kozeny's theory, which gave the groups f and Re', assumed that the packing can be treated as a series of tortuous passages; we should not, therefore, expect these groups to be appropriate for very high voidages.

Table 1. *Comparison between the result of Richardson and Zaki* (1954) $U/U_t = \epsilon^n$ *and expressions derived from fixed-bed data*

	ϵ	0·3	0·4	0·5	0·6	0·7	0·8	0·9	1·0
Reynolds number < 0·2	$\epsilon^{4·65} \times 100$	0·373	1·41	4·0	9·3	19·1	35·4	61·1	100
	$\dfrac{3\epsilon^3 \times 100}{25(1-\epsilon)}$	0·463	1·28	3·0	6·46	13·7	30·7	87·4	∞
High Reynolds number	$\epsilon^{2·39} \times 100$	5·68	11·2	19·1	29·5	42·7	58·6	77·7	100
	$0·434\epsilon^{3/2} \times 100$	7·13	11·0	15·4	20·2	25·4	31·1	37·1	43·4

(ii) For the medium voidages (0·5–0·8) in table 1, the values of U/U_t are systematically higher for fluidised than for fixed beds; this means that the fluid experiences a lower resistance in a fluidised bed than in a fixed bed, because a higher velocity is necessary to support the weight of the particles. A similar effect was noted by Hawksley (1951, p. 133), by Loeffler and Ruth (1959) and by Richardson and Meikle (1961). Happel and Brenner (1957) and Adler and Happel (1962) have suggested that this lower resistance of a fluidised bed is due to slow internal circulation of the particles within the bed, the particles moving up in the middle and down near the containing walls. But it may be that there is a more fundamental difference between a particulately fluidised bed and a fixed bed having the same voidage, since in the former case the particles are free to oscillate and to rotate; such movements would certainly affect the fluid flow but would probably increase rather than decrease the resistance. Such particle oscillations, if they occur, would be very difficult to distinguish from the effects of bubbling in an aggregatively fluidised bed, and certainly a bed of 0·1 mm glass beads, for example, does not seem to exhibit particle movements when fluidised by water. Happel's suggestion of a slow circulation of particles within the whole bed is therefore the most likely explanation of the lower resistance to flow.

Finally, it must be remembered that (1.20) can only be applied to fluidised beds that are particulate, or non-bubbling. Chapter 5 gives a method of deciding whether any particular fluidised system will contain bubbles of appreciable size; this criterion should be considered before using (1.20) to calculate the expansion of a fluidised bed.

1.7. The two-phase theory of fluidisation

A model of aggregative fluidisation may be set up by considering a bed as a two-phase system consisting of

(a) a particulate phase in which the flow-rate is equal to the flow-rate for incipient fluidisation, i.e. the voidage fraction is essentially constant at ϵ_0, and

(b) a bubble phase which carries the additional flow of fluidising fluid.

The evidence for this picture of a fluidised bed is considerable. It finds support, for instance, in the work and differing techniques of Morse and Ballou (1951) (capacitance probe), Toomey and Johnstone (1952) (γ-ray method), Yasui and Johanson (1958) (light transmission), Dotson (1959) (capacitance probe), Baumgarten and Pigford (1960) (visual observation) and Romero and Johanson (1962). On the other hand, the assumption that all the excess fluid above that required for incipient fluidisation passes through a fluidised bed in the bubble phase has been challenged by Lanneau (1960), but it is possible that his experimental technique failed to detect bubbles below a certain size. It is also noteworthy that his conclusions are primarily based on experiments carried out at superficial fluidising velocities of more than $100U_0$. This underlines the fact that the two-phase model of a fluidised bed has so far come largely from work on systems with fluidising velocities of less than about $10U_0$.

In this book some of the consequences of assuming that the particulate phase has the voidage at incipient fluidisation, and that the bubble phase carries the extra flow, are worked out from first principles. This theory provides evidence for the two-phase model of fluidisation in so far as it agrees with experiment. Nevertheless, it is necessary to emphasise that the theory appears to be reasonable only for fluidised systems whose behaviour is largely characterised by random bubbles. Thus, for instance, the theory is not directly applicable to beds exhibiting extensive channelling, spouting, or pneumatic transport.

Chapters 2 and 3 examine the properties of bubbles in fluidised beds—such as their formation and rising velocity—and show how the bubbles behave as if they were in a liquid like water but of small surface tension. Chapter 4 gives an analysis of the flow of fluidising fluid in the neighbourhood of a rising bubble; this leads to an explanation of why the bubble manages to behave as if it were in an inviscid liquid, and to an estimate of the exchange of fluid between the bubble and the surrounding particles. Chapter 5 gives an analysis of bubble stability which provides a link between the apparently different phenomena of aggregative and particulate fluidisation. In Chapter 6 the bubble model is applied to the analysis of what is perhaps the most important practical application of fluidised beds, the fluidised catalytic reactor. Throughout the book the physical principles and ideas are related as closely as possible to experiment.

CHAPTER 2

THE RISE AND COALESCENCE OF BUBBLES IN FLUIDISED BEDS

2.1. Introduction

The bubbles of air present in a conventional air–solids fluidised bed have two properties which are readily apparent: they rise at a finite velocity and, in general, the bubbles grow in size as they move up through the bed. These properties are the subject of this chapter.

It needs to be stressed that the information that has been collected to date is not complete. For example, measurements of the rising velocities of bubbles have largely been confined to bubbles essentially uninfluenced by nearby bubbles of a similar kind; and also, bubble coalescence has so far only been studied for bubbles one above the other in a vertical chain. Nevertheless, despite these limitations, a striking similarity has been found between the behaviour of large gas bubbles in liquids and bubbles in fluidised beds, and this similarity leads to predictions of a useful and quantitative kind. Furthermore, it provides an insight into the fluid mechanical principles that govern a fluidised system.

Before considering the behaviour of bubbles in fluidised beds, we shall give an account of the relevant properties of bubbles in ordinary liquids.

2.2. Bubbles and drops in liquids

In a liquid of small viscosity, the rate of rise of a large bubble is governed by inertia forces, surface tension and viscous effects being negligible by comparison. The theoretical problem of calculating the rate of rise of the bubble is complex, however, because the shape of the bubble is determined by the flow outside it, the shape adjusting itself so that the pressure within the bubble is constant, the density of the gas within the bubble being small.

(a) A long bubble in a tube

Fig. 6 indicates the flow pattern in the case that is easiest to analyse, that of a very long bubble rising in a tube. The bubble is imagined to be held stationary by a downward flow of liquid and the pressure on the surface of the bubble $ROPR'$ must be constant. Therefore on applying Bernoulli's theorem between the stagnation point O and a typical point P, the pressure at both points being the same, we get

$$w_s = (2gz)^{\frac{1}{2}}, \qquad (2.1)$$

where w_s is the liquid velocity at P, z is the vertical distance from O to P, and g is the acceleration of gravity. Equation (2.1), together with the condition that the velocity U_b shall be uniform across AA', far above the bubble, constitute the boundary conditions for the problem. The flow of liquid within the region $AA'B'R'PORBA$ is governed by the equation of potential flow, (A. 11) in Appendix A (p. 127); the shape of $ROPR'$ must be adjusted so that (2.1) is satisfied. Dumitrescu (1943), and later Davies and Taylor (1950), gave approximate solutions, the former giving the result

$$U_b = 0.35(gD)^{\frac{1}{2}}, \qquad (2.2)$$

where D is the tube diameter. From (2.1) it is clear that the relation between U_b and D must be of this form and Davies and Taylor gave a slightly different value of the empirical constant; in both papers the theoretical result was supported by experiments to measure the rising velocity of a long bubble in a tube. These experiments are extremely easy to perform; the reader can verify (2.2) by taking a long vertical tube corked at both ends and filled with water; when the bottom cork is suddenly removed, a bubble of the type shown in fig. 6 is readily observed to rise through the water.

Nicklin, Wilkes and Davidson (1962) have recently shown that (2.2) applies to a finite bubble or 'slug' rising in a tube. This result might be anticipated from the theory, because the rising velocity of the bubble is determined primarily by the flow near its nose; if the height of the bubble is more than one or two tube diameters, the wake below the bubble does not influence the nose and hence has no effect upon the rising velocity.

(b) A bubble rising in a large volume of liquid

The same sort of theory can be used when the tube diameter is much larger than the bubble. Fig. 7 shows the flow pattern, the bubble OQQ' again being imagined to be held stationary by a downward flow of liquid. Below QQ' is a stagnant wake $QQ'R'R$ (shown hatched). Outside the curved wake and bubble boundary $RQOQ'R'$, the liquid moves downwards in such a way as to satisfy the following boundary conditions, derived from Bernoulli's theorem:

between O and QQ', $\qquad w_s = (2gz)^{\frac{1}{2}}$ $\quad (0 < z < h)$, $\quad$ (2.3)

between QQ' and RR', $\quad w_s = (2gh)^{\frac{1}{2}}$ $\quad (z > h)$, $\qquad$ (2.4)

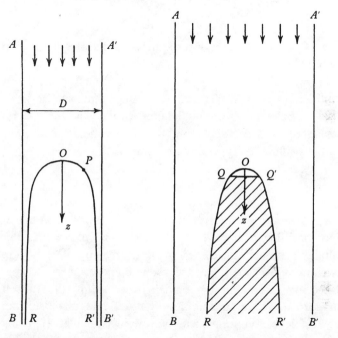

Fig. 6. Wakeless bubble held stationary in a vertical tube.

Fig. 7. Flow of ideal fluid (zero viscosity) round a spherical-cap bubble (Rippin, 1959).

where h is the bubble height, from O to QQ'. Equation (2.3) is derived from the fact that the pressure within the bubble is constant, and (2.4) from the fact that the pressure within the wake increases linearly with height, the wake liquid being stagnant. Of course the flow pattern depicted in fig. 7 involves a sudden change in velocity across the curved surface of the wake; but we are here thinking of a fluid with truly zero viscosity for which such discontinuities are possible. In a real fluid, the wake is turbulent, although the 'free streamline' QR can be identified a short distance below the bubble.

To solve the ideal fluid problem, it is necessary to find the shape $RQOQ'R'$ such that the boundary conditions (2.3) and (2.4) are satisfied by irrotational flow within the region $AA'B'R'Q'OQRBA$. Rippin (1959) gave a numerical solution to this problem, using a computer to solve (A. 12) (p. 127) and adjusting the shape of the wake and bubble boundaries to satisfy (2.3) and (2.4). Rippin's calculations gave the rising velocity of the bubble,

$$U_b = 0.914(gD_e)^{\frac{1}{2}}, \qquad (2.5)$$

where D_e is the diameter of the sphere that has the same volume as the bubble.

As with the long bubble in a tube, it is clear from the boundary conditions (2.3) and (2.4) that the form of (2.5) must be as shown and it only remains to determine the numerical coefficient. Davies and Taylor (1950) gave a value found from experimental observations of bubble rising velocity, and their result is

$$U_b = 0.711(gD_e)^{\frac{1}{2}}. \qquad (2.6)$$

The result was also based on the experimental observation that the flow near the front of a spherical-cap bubble is very similar to the theoretical flow near the front of a complete sphere in an inviscid liquid. This conclusion came from measurements of (i) the pressure distribution over the surface of a solid body of spherical-cap shape placed in a steady air flow, and (ii) the rate of rise and radii of curvature of large air bubbles in water and nitrobenzene; the conclusion is also supported by Rippin's calculations using the model of fig. 7.

At this point it is worth contrasting the approaches made by Davies and Taylor and by Rippin. An exact solution to an idealised problem is obtained by Rippin by supposing the wake of the bubble to be stagnant; whereas Davies and Taylor show that, although the actual fluid motion in the wake cannot be precisely described, the upward motion of the bubble is largely determined by the fluid flow near its nose. The theory of Davies and Taylor is given here, since it is important in interpreting some of the results from fluidised systems. They imagined a spherical-cap bubble, shown in fig. 8, held fixed by a down flow of liquid, and assumed that the flow near the front of the bubble is the same as in potential flow round a sphere of radius R, the radius of the spherical cap. The relevant velocity potential is given by (A. 15) in Appendix A (p. 128), with R in place of b, and the surface velocity

$$w_s = \left(\frac{1}{r}\frac{\partial \phi}{\partial \theta}\right)_{r=R} = -\tfrac{3}{2}W\sin\theta. \tag{2.7}$$

Now $-W$, the velocity at infinity, must equal U_b the rising velocity of the bubble, and to satisfy the boundary condition (2.1) we must have

$$2gz = 2gR(1-\cos\theta) = \tfrac{9}{4}U_b^2\sin^2\theta,$$

θ now being equivalent to α in fig. 8.

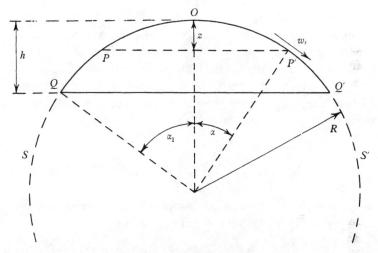

Fig. 8. Flow at the front of a spherical-cap bubble predicted from ideal flow round a sphere.

This equation can be satisfied at one value of θ or for all small values of θ, and taking the latter condition, we put $\theta \to 0$ and $\sin^2\theta/(1-\cos\theta) \to 2$, giving

$$U_b = \tfrac{2}{3}(gR)^{\frac{1}{2}}. \tag{2.8}$$

Davies and Taylor showed experimentally, by measuring the rate of rise and radius of curvature of a given bubble, that (2.8) relates U_b and R very well for large air bubbles. Their photographs also show that the boundary between the wake below the bubble and the flowing stream outside it can be identified between QQ' and SS', fig. 8. Below point S, turbulent mixing, between the wake and the stream outside, blurs the wake boundary though the turbulence is roughly confined within the sphere of radius R. It is this turbulence in the wake which gives the discrepancy between Rippin's theoretical result (2.5) and the experimental result (2.6) of Davies and Taylor; Rippin showed that the difference is due to the fact that the increase of pressure with distance below the bubble is somewhat greater than the increase of pressure due to the static head of liquid postulated in the model of fig. 7; this is further discussed in Chapter 4 and Appendix B.

(c) The rising velocity of large drops

This section deals with the application of the Davies and Taylor equation (2.6) to the case of a large drop of liquid rising in another liquid. Interfacial tension is neglected and it is assumed that the liquid within the drop is stagnant, so the pressure varies linearly with height instead of being constant as it is within a gas bubble. Hence, in fig. 8, by applying Bernoulli's theorem between O and PP' to the flowing liquid just outside the drop we get

$$w_s^2 = 2gz(1 - \rho_f/\rho_c), \tag{2.9}$$

where ρ_f is the density of the drop and ρ_c is the density of the continuous phase. Equation (2.9) is similar to (2.3), but with the multiplying factor $(1 - \rho_f/\rho_c)$, and it can therefore be inferred from (2.6) that the rising velocity of a drop should be

$$U_b = 0.711[gD_e(1 - \rho_f/\rho_c)]^{\frac{1}{2}}. \tag{2.10}$$

A criticism of this result is that it ignores internal circulation within the drop; nevertheless (2.10) is in good agreement with the results of Klee and Treybal (1956) for large drops for which the effects of interfacial tension and viscosity are small. The comparison between theory and experiment is shown in table 2. Moreover, Harmathy (1960) has recently shown by dimensional analysis that the bubble-rising velocity is proportional to $(\rho_c - \rho_f)^{\frac{1}{2}}$ in agreement with (2.10).

(d) The rising velocity of continuously generated bubbles

When a stream of bubbles in a vertical tube is generated continuously by blowing air in at the bottom, the absolute upward velocity of each bubble is greater than the velocity with which the same bubble would rise in stagnant liquid.

Table 2. *Rising velocities of drops in liquids*
(*Klee and Treybal, 1956*)

System no.	Equivalent drop diam. D_e (cm)	ρ_f (g/ml)	ρ_c (g/ml)	Rising velocity U_b (cm/s) Experimental	Rising velocity U_b (cm/s) Equation (2.10)
7	0·5	0·837	0·960	5·5	5·6
6	0·6	1·011	1·145	6·7	5·9
5	0·6	0·866	0·9705	6·0	5·7
4	0·8	0·8242	0·9982	9·1	8·3
11	0·7	0·9200	0·9980	7·8	5·2
9	0·9	0·8155	0·9947	10·3	9·0

A particular case of this kind is shown in fig. 9, following Griffith and Wallis (1961) and Nicklin *et al.* (1962); a vertical tube of cross-sectional area A contains water through which air is blown steadily at a volume flow rate G. The bubble pattern in the lower part of the tube depends upon the inlet conditions, but further up the tube a unique pattern of regular bubbles, known as 'slug flow', is formed. Consider now a fixed control surface $CC'D'D$ enveloping the lower part of the tube and cutting through the clear liquid between the bubbles at section CC' in fig. 9. During a small time dt, chosen so that no bubbles cross the section CC', a volume of gas $G\,\mathrm{d}t$ must enter at the bottom, and a corresponding volume of liquid must pass across CC', and therefore the mean liquid velocity across CC' must be G/A, neglecting the compressibility of the gas. Of course the liquid velocity across CC', averaged over a long period, must be zero, but when a bubble crosses CC' there is a downward flow which compensates for the upward velocity G/A at the instant shown in

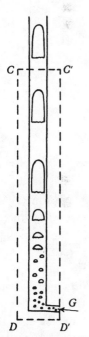

Fig. 9. Continuously generated bubbles in 'slug flow'.

fig. 9. Now comparing figs. 6 and 9, the relative velocity between the bubble in fig. 6 and the liquid at section AA' is $U_b = 0·35(gD)^{\frac{1}{2}}$, the natural rising velocity of the bubble in still water. In fig. 9 we should therefore expect the *relative velocity* between the liquid at CC' and the bubble below to be $0.35(gD)^{\frac{1}{2}}$, and hence its *absolute velocity* should be $G/A + 0.35(gD)^{\frac{1}{2}}$. Nicklin *et al.* (1962) found from experiments that the absolute bubble velocity is

$$1.2G/A + U_b. \tag{2.11}$$

The factor 1.2 is necessary because of the non-uniform velocity profile at section CC' where the liquid is in turbulent flow for which the peak velocity at the middle of the tube is about 1.2 times the mean velocity; the bubbles evidently rise relative to the fastest moving liquid in the middle of the tube.

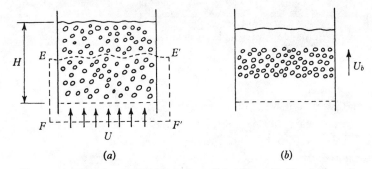

Fig. 10. (*a*) Continuously generated small bubbles. (*b*) Cloud of small bubbles rising in a stagnant liquid (Nicklin, 1961).

Nicklin (1962) showed that a similar situation occurs when a stream of small bubbles is continuously generated in a large tube, as shown in fig. 10 (*a*). The liquid is supported on a porous grid through which air is blown continuously to form small bubbles, and the above argument can be applied in this case. By considering the flow into the control surface $EE'F'F$, the surface EE' passing through the liquid between the bubbles, it can be seen that this liquid must have the upward velocity U of the gas. Considering now fig. 10 (*b*), in which a cloud of bubbles each of the same bubble size as in fig. 10 (*a*) is rising through stagnant liquid with a velocity U_b, we see that the absolute velocity of the bubbles in fig. 10 (*a*) must be

$$U_A = U + U_b. \tag{2.12}$$

Nicklin (1962) has verified this result by experiments with air and water, the cloud of bubbles in fig. 10 (*b*) being generated by turning the air flow suddenly on and then off again. Wallis (1962) has shown how (2.12) can be applied to a wide variety of two-phase systems.

Nicklin (1962) has also shown how the bubble velocity U_b is related to the rise of the liquid level caused by the bubbling. If H is the liquid height when the gas velocity is U, H_0 is the height when $U = 0$, and if in unit volume there are N bubbles each of volume V, then for continuity of the gas flow,

$$NVU_A = U, \qquad (2.13)$$

and from the fact that the expansion $H - H_0$ is due to the bubbles within the liquid,

$$H - H_0 = NVH. \qquad (2.14)$$

From (2.12), (2.13) and (2.14), eliminating NV,

$$U_A/H = U_b/H_0, \qquad (2.15)$$

and

$$U_b/U = H_0/(H - H_0). \qquad (2.16)$$

An equation similar to (2.16) has been verified for air–water by Nicklin (1962). The equation provides a means of estimating U_b from observable quantities.

It may be noted here that (2.12) to (2.16) are applicable to a bubbling fluidised bed with U replaced by $U - U_0$, if the assumption is made that the additional flow above the amount for incipient fluidisation passes through the bed as bubbles. To a first approximation, it can be assumed that U_b is the same as for an isolated bubble, and hence *for a fluidised bed*, from (2.6) and (2.16)

$$\frac{0 \cdot 711 (g D_e)^{\frac{1}{2}}}{U - U_0} = \frac{H_0}{H - H_0}, \qquad (2.17)$$

where H_0 denotes the bed height at incipient fluidisation. This has been used by Orcutt, Davidson and Pigford (1962) to estimate the equivalent bubble diameter D_e from observations of bed height H as a function of velocity U for a bed of catalyst particles fluidised by air.

2.3. The rate of rise of bubbles in fluidised beds

(a) Bubble shape

The direct evidence concerning the mode of rise of bubbles in fluidised beds is that when bubbles reach a bed surface they erupt as a spherical cap, as illustrated in Plate I. A bubble therefore appears to rise through the bed by laterally displacing solid particles around its upper surface rather than as a result of the collapse of that surface. That the bubble shape is similar to that of a large air bubble in water is shown by Plate II which is an X-ray photograph of an air bubble rising in an incipiently fluidised bed of glass Ballotini (Rowe, 1962 b, see also Rowe, Partridge, Lyall and Ardran, 1962). Plate II shows the characteristic spherical top which is observed for a large air bubble in water; but the included angle α_1 is different, being about 120° in Plate II compared with 50° for air–water.

All other evidence on the shape of bubbles in fluidised beds is indirect, namely:

(i)   Photographs of bubbles in 'two-dimensional' beds (Plate III, Rowe, 1962 a; Rowe and Partridge, 1962).

(ii)   Photographs of bubbles taken near the walls of fluidised beds (Harrison, 1959; Zenz and Othmer, 1960, p. 266).

(iii)   The investigation of bed voidage by light probe or radiation techniques (Yasui and Johanson, 1958; Baumgarten and Pigford, 1960).

(iv)   Information on bubble shape can be deduced from measurements of bubble velocity (§ 2.3 (f), p. 38).

(b)   The steady rate of rise of bubbles in fluidised beds

Four groups of workers have so far been concerned to measure the rising velocities of bubbles in fluidised beds. Baumgarten and Pigford (1960) used a γ-ray absorption technique to measure the thickness and frequency of bubbles in the bed, and thereby obtained an indirect measure of the bubble velocity. Yasui and Johanson (1958) measured the rising velocity directly using a light probe. It is not, however, possible to make a clear comparison between this work and the measurements of Davidson, Paul, Smith and Duxbury (1959) and Harrison and Leung (1962 a) on the rising velocities of *single* gas bubbles injected into an essentially quiescent bed, fluidised slightly above the incipient condition.

The effect of the proximity of other bubbles on the rising velocity of a given bubble has yet to be studied quantitatively.

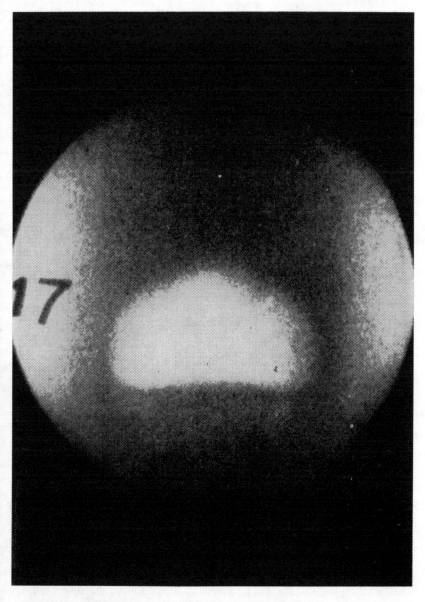

Plate II. X-ray photograph of a bubble (about 2 in diameter) rising in a 6 in diameter incipiently fluidised bed of glass beads (o·5 mm diam.). $U_0 \simeq$ 45 cm/sec. $U_b \simeq$ 50 cm/sec (Rowe, 1962 b).

(*Facing p.* 30)

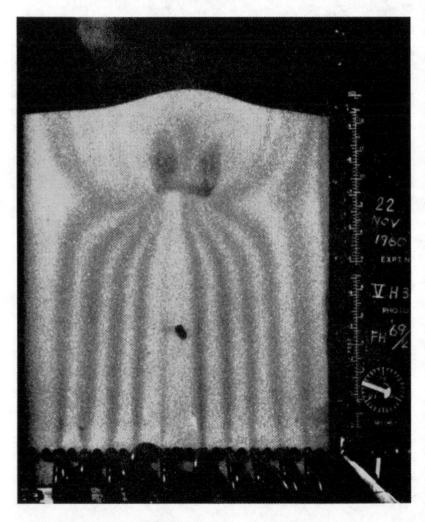

Plate III. Photograph of a 'two-dimensional' bubble within an air-fluidised bed between two flat plates. The traces are injected NO_2 (Rowe, 1962a).

(c) The rising velocity of a single bubble in a quiescent bed

(i) *Experimental*. The rate of rise of air bubbles in an air-fluidised bed was measured (Davidson *et al.* 1959; Harrison and Leung, 1962 a) by observing the time interval between the injection of the bubble of known volume at the base of the bed and its arrival at the bed surface. The first experiments (Davidson *et al.* 1959) used columns of about 3 and 6 in diameter, and the volumes of bubbles studied ranged from 5·8 to 389 ml. The experiments were carried out in turn with particles of glass (0·015 cm mean diameter), sand (0·04 cm), and swede seeds (0·17 cm).

In a further series of experiments (Harrison and Leung, 1962 a), an apparatus 2 ft square in cross-section was used in order to measure the rising velocity of bubbles ranging in volume from 25 to 10,000 ml. In this work an air bubble was injected into a quiescent bed of sand particles.

Results from these experiments are shown in figs. 11 and 12 where the measured time of rise is plotted as a function of the depth of injection. The most striking feature is that each set of points lies nearly on a straight line, showing that the velocity of the bubble, given by the slope of the line, does not appear to vary as the bubble rises. Another feature of the results is that the best straight line through each set of points does not pass through the origin. The positive intercepts on the abscissae can be explained if the injected bubble was given an initial velocity rather larger than the steady rising velocity. It is reasonable that this should be so since the bubbles were injected from a pressure vessel at a pressure in excess of that in the bed. The intercept defines an 'entrance effect', L_0, which is a function of the volume injected.

(ii) *Analysis of results*. In § 2.2 it was shown that the ratio $U_b/V^{\frac{1}{6}}$ is sensibly constant for spherical-cap gas bubbles in liquids. Davidson *et al.* (1959) suggested that spherical-cap bubbles in fluidised beds might be described by an equation of the same form, and table 3 shows that $U_b/V^{\frac{1}{6}}$ is in fact approximately constant for fluidised systems.

The analysis of the spherical-cap bubble given in § 2.2 makes no allowance for wall effect. Uno and Kintner (1956) measured the rising velocities of gas bubbles in liquids contained in tubes of various diameters, and compared their results with the bubble's rising velocity $U_{b\infty}$ in an infinite medium. They plotted the ratio $U_b/U_{b\infty}$ as a function either of $1-(D_e/D)$ or of $1-(D_f/D)$, where D_f is the frontal diameter of the bubble, and these results are summarised in fig. 13, from which it is seen that the wall effect becomes negligible when D_e is less than about one-tenth of D. Fig. 13 also shows $U_b/U_{b\infty}$ obtained from the ratio of (2.2) to (2.6); this shows that there is slug flow—the wall effect predominating—when D_e is greater than about one-third of D.

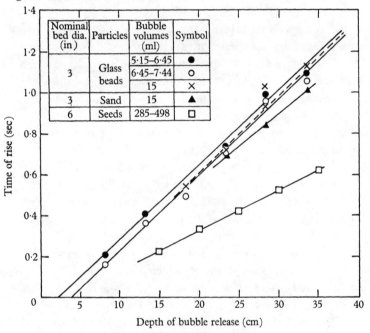

Nominal bed dia. (in)	Particles	Bubble volumes (ml)	Symbol
3	Glass beads	5·15–6·45	●
		6·45–7·44	○
		15	×
3	Sand	15	▲
6	Seeds	285–498	□

Fig. 11. Single bubble rising in an incipiently fluidised bed
(Davidson *et al.* 1959).

Table 3 shows the results of applying Uno and Kintner's correction $m = U_b/U_{b\infty}$ for a gas–liquid system to the rising velocities of bubbles in fluidised beds. Davidson *et al.* (1959) used D_e/D in fig. 13 because they did not observe D_f; Harrison and Leung (1962a) used their observed values of D_f; both sets of authors were thus able to obtain values of $U_{b\infty}/V^{\frac{1}{6}}$, and table 3 shows that these values are in reasonable agreement with the value of 24·8 cm$^{\frac{1}{2}}$ sec^{-1} observed by Davies and Taylor for air bubbles in inviscid liquids.

Fig. 14 shows a graphical summary of the results of Harrison and Leung, m being the correction term of Uno and Kintner, and t_s the time taken for the bubble to reach the surface. The least square line through the origin and the 229 experimental points is $H - L_0 = 22 \cdot 3 m V^{\frac{1}{6}} t_s$, so that

$$U_{b\infty}/V^{\frac{1}{6}} = 22 \cdot 3 \, \text{cm}^{\frac{1}{2}} \sec^{-1}, \tag{2.18}$$

and
$$U_{b\infty} = 0 \cdot 71 g^{\frac{1}{2}} V^{\frac{1}{6}}. \tag{2.19}$$

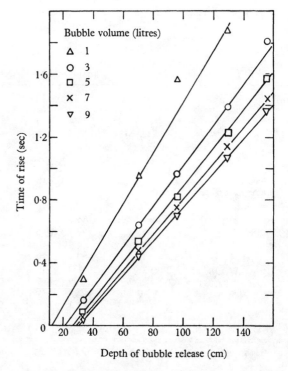

Fig. 12. Single bubble rising in an incipiently fluidised bed (Harrison and Leung, 1962 a).

The standard deviation of the points from the least-square line is 0·11; there is more scatter in the results at higher values of $(H - L_0)/V^{\frac{1}{6}}$, which relate to a deep bed and small bubbles. The value of 22·3 in (2.18) may be compared with the average value of 22·8 of Davidson et al. (excluding their extrapolated result for a 389 ml bubble), and with the value of Davies and Taylor of 24·8.

It thus appears that a bubble in a gas-fluidised bed rises slower than it would in an inviscid liquid. But in subsequent chapters in this book, bubble rising velocities are calculated using the coefficient of 24·8 of Davies and Taylor rather than the value of 22·3 given in (2.18); it was felt that the difference between the coefficients was not significant for the purposes in hand.

Table 3. *Velocities of bubbles in fluidised beds*

Ref.	Bed size	Material	Approx. diam.	V (ml)	U_b (cm s^{-1})	$U_b/V^{\frac{1}{6}}$ (cm$^{\frac{1}{2}}$s^{-1})	$U_{b\infty}/V^{\frac{1}{6}}$ (cm$^{\frac{1}{2}}$s^{-1})
Davidson *et al.* (1959)	7·53 cm diam. ($\simeq$ 3 in)	Glass beads	0·015 cm	5·80	27·7	20·7	23·4
				7·09	27·0	19·5	22·4
				15·0	26·0	16·6	21·2
		Sand	40–60 mesh, 0·04 cm mean	15·0	29·5	18·8	24·2
	14·6 cm, diam. ($\simeq$ 6 in)	Swede seeds	1·7 mm mean	389	50	18·6	31·1*
Harrison and Leung (1962 a)	2 ft square	Sand	Mostly 72–120 mesh, i.e. approx. 0·015– 0·02 cm	1,000	60·9	19·3	20·6
				2,000	67·7	19·1	20·9
				3,000	76·9	20·2	22·7
				4,000	80·9	20·3	23·3
				5,000	81·9	19·8	23·0
				6,000	84·9	19·9	23·4
				7,000	88·0	20·1	24·1
				8,000	90·3	20·2	24·4
				9,000	92·3	20·2	24·8
				10,000	93·8	20·2	24·9

* Based on an extrapolated correction for $U_b/U_{b\infty}$.

(d) The use of the Davies and Taylor equation for bubbles in gas-fluidised beds

The Davies and Taylor equation contains the one-sixth power of the bubble volume and is thus insensitive with respect to that parameter. This, together with the fact that an experimental error of some 10 % must be expected from the measurements of bubble velocity, means that it is difficult to substantiate the equation reliably for fluidised systems, though the difficulty is no greater than for an ordinary liquid. The above application of the Davies and Taylor equation is evidence for the liquid-like character of gas-fluidised beds, but it is by no means the only evidence as may be seen in Chapter 3. The use of (2.19) must be viewed against this background of evidence, for when the equation is taken strictly by

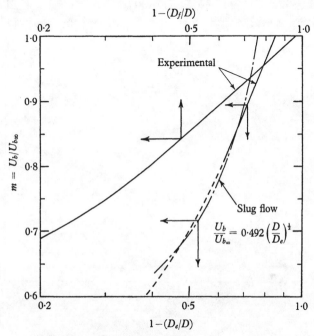

Fig. 13. Bubble rising velocity U_b in a tube of diameter D
(Uno and Kintner, 1956).

itself, its insensitivity makes it a rather slender pointer to the 'two-phase' nature of gas-fluidised beds. The following detailed comments on the application of (2.19) can be made:

(i) The bubble's rising velocity may be expected from equation (2.19) to be independent of the type of solid particle making up the bed; and this was confirmed by Yasui and Johanson (1958) who found U_b independent of the density and size of the particles, and the gas flow to the bed. All the results from the experiments described in § 2.3 (c) (i) (p. 31) using glass, sand and swede seeds can also be represented together.

(ii) If there is substantial leakage from the bubble as it is formed at the injector then the assumption made in §2.3 (c) (ii) (p. 31) that the volume of the bubble was equal to the injected volume would be erroneous. Davidson *et al.* (1959) compared the volume of the suddenly injected bubble with the volume estimated from the rise of the surface of the bed, and found that these quantities agreed within the limits of experimental error (i.e. better than 10 %). There is also other evidence, discussed in Chapter 3, that there is little

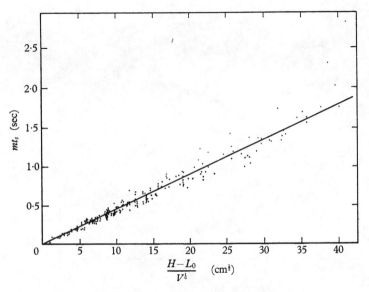

Fig. 14. Fluidised bed 2 ft square. Data on the time of rise of bubbles (Harrison and Leung, 1962 a).

leakage from the bubble as it is formed in a bed of particles of about 0·01 cm diameter, although the leakage may possibly be greater with larger particles.

(iii) When the diameter of the bubble approaches that of the containing vessel the bubble is elongated and its velocity is less than that predicted by (2.19). It is shown in §2.3 (c) (ii) (p. 31) that it is possible to make some estimate of this 'wall effect' by using the data of Uno and Kintner (1956). When the bubble is so large as to become a 'slug', its rate of rise can be predicted from the result of Dumitrescu (1943), and this is considered more fully in §2.4 (p. 41).

(iv) Equation (2.19) will clearly be inappropriate for the régime of fluid–solids behaviour between the so-called 'slugging' region and the point of pneumatic transport. Furthermore, a description of fluidisation in terms of the properties of bubbles will, of necessity, be inadequate for systems in which the fluid largely 'channels' or 'spouts' through the bed.

(e) The use of the Davies and Taylor equation for bubbles in liquid-fluidised beds

Unlike many liquid-fluidised beds, the presence of water 'bubbles' is a feature of the appearance of a bed of lead shot fluidised by water. From what has been said about bubbles in a gas-fluidised bed, it might be expected that (2.10), for liquid–liquid systems, would be applicable to the rise of a liquid-filled void in a liquid-fluidised bed.

The density of the bubble phase is ρ_f, and at incipient fluidisation the density of the particulate phase is $[\rho_s(1-\epsilon_0)+\rho_f\epsilon_0]$. Substituting these densities in (2.10), but using the coefficient from (2.19), we get

$$U_b = 0.71 g^{\frac{1}{2}} V^{\frac{1}{6}} \left[(1-\epsilon_0) \Big/ \left(\frac{\rho_s}{\rho_s-\rho_f} - \epsilon_0 \right) \right]^{\frac{1}{2}}. \qquad (2.20)$$

An attempt has been made (de Kock, 1961) to measure the rising velocities of water bubbles in a water-fluidised bed of lead shot. This investigation proved somewhat inconclusive for it was attended by two major difficulties. First, it was not easy to be certain as to the volume of the bubble, for some water appeared to be lost on injection, and also the rising bubbles were not entirely stable and tended to break up. This latter observation is further considered in Chapter 5. A second difficulty was that due to the high density of the system studied it did not prove practicable to use beds greater than about 15 cm diameter, and this limited the range over which the bubble volume could be varied. However, results that were obtained from these experiments were not inconsistent with (2.20).

(f) *The relation between the rising velocity of a*
spherical-cap bubble and its shape

For the spherical-cap bubble OQQ' shown in fig. 8, the relation between the bubble volume V and R and α_1 can be obtained by integrating from $\alpha = 0$ to α_1, giving

$$V = \pi R^3[\tfrac{2}{3} - \cos\alpha_1 + \tfrac{1}{3}\cos^3\alpha_1]. \tag{2.21}$$

Hence, using (2.8), $U_b = cg^{\frac{1}{2}}V^{\frac{1}{6}},$ (2.22)

where $c = \tfrac{2}{3}\pi^{-\frac{1}{6}}(\tfrac{2}{3} - \cos\alpha_1 + \tfrac{1}{3}\cos^3\alpha_1)^{-\frac{1}{3}}.$ (2.23)

By measuring U_b, Davies and Taylor found $c = 0.792$ [cf (2.6)] for gas–liquid systems. Equation (2.23) then gives $\alpha_1 = 50°$. This value for the included angle of the bubble compared well with the mean angle given by visual observation, although the scatter of the results showed clearly that all the bubbles were not exactly similar.

For a gas-fluidised bed $c = 0.71$, from (2.19), and this corresponds to $\alpha_1 = 60°$; which is consistent with the evidence listed in § 2.3 (a) (p. 30) that bubbles in fluidised beds are closer in shape to the hemispherical than bubbles in an ordinary two-phase system. However, the value of $\alpha_1 = 60°$ is not in agreement with the value found by X-ray photography, and a possible explanation of this discrepancy is discussed in § 2.3 (g) (p. 40).

It is also possible to estimate α_1 by measuring the maximum diameter, D_f, of a bubble (of known volume V) as it breaks the surface of the bed. D_f should be proportional to $V^{\frac{1}{3}}$ if the bubbles are geometrically similar, and this has been verified in an apparatus 2 ft square for bubbles of less than 16 l. Fig. 15 demonstrates this proportionality, together with the fact that it does not hold for diameters of bubble approaching that of the column. The equation of the best straight line relating D_f and $V^{\frac{1}{3}}$ is

$$D_f = 1.92V^{\frac{1}{3}}. \tag{2.24}$$

To calculate α_1 from this result, V is eliminated from between (2.21) and (2.24), and then substituting $D_f = 2R\sin\alpha_1$ gives $\alpha_1 = 64°$. It may be that the difference between the values of α_1 derived from (2.24) (64°) and (2.19) (60°) arises because D_f was determined from the disturbance on the bed surface caused by the erupting bubble, and this is likely to be somewhat greater in size than the bubble itself. Recent observations have cast some doubt on the method of determining the volume of a bubble from the measurement of its frontal diameter as it breaks surface. Bubbles in 'two-dimensional' fluidised beds appear to expand considerably just before surfacing, and thus the frontal diameter observed at the top of the bed is rather greater than the frontal diameter of the bubble over most of its passage through the bed. It is not yet clear whether this sudden

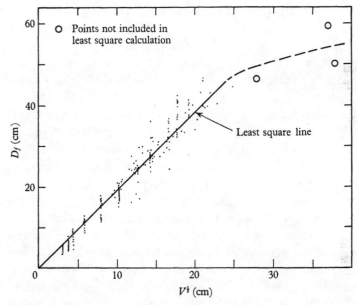

Fig. 15. The frontal diameter of a bubble as a function of (bubble volume)$^{\frac{1}{3}}$ (Harrison and Leung, 1962a).

expansion occurs with bubbles in normal three-dimensional apparatus, or whether it arises from the influence of the walls in 'two-dimensional' beds.

This indirect evidence on the shape of bubbles indicates a smaller included angle than is suggested by photographs of bubbles near the walls of fluidised beds (Harrison, 1959), or in 'two-dimensional' apparatus (Plate III, Rowe, 1962a). In these cases it is not known to what extent the walls distort the shape of the bubble. But Rowe's X-ray photograph of a single bubble rising in an incipiently fluidised bed (Plate II) shows that the angle α_1 is of the order of 120° which is also higher than the above values deduced from the rising velocity and the diameter of the surfacing bubble. All the values of α_1 for fluidised beds are greater than the value (50°) for an air bubble in water. It is possible that these differences are due to the high 'viscosity' of the fluid–solid state, and this will now be considered.

(g) Viscosity of a fluidised bed

A number of investigators have measured the apparent viscosity of a fluidised bed by using methods available for ordinary liquids. Thus Matheson, Herbst and Holt (1949) and Furukawa and Ohmae (1958) used a rotating paddle viscometer, and observed viscosities of the order of 0·5 to 20 P (Poise) for a fully fluidised bed, the viscosity increasing with the size of the particles forming the bed. The viscosity was much higher at incipient fluidisation. Similar results were obtained by Kramers (1951), who used a different kind of paddle, and by Diekman and Forsythe (1953) who immersed in a fluidised bed a gauze cylinder which rotated about its vertical axis. Schügerl, Merz and Fetting (1961) have published very comprehensive results obtained by containing a gas-fluidised bed within the annulus between two concentric cylinders with a vertical axis. They measured the torque necessary for relative rotation of the cylinders about the axis, and obtained results that are in remarkable agreement with the earlier results from the paddle viscometers. For small shear stresses the fully fluidised bed was found to behave as a Newtonian fluid, although once again the viscosity was very high at incipient fluidisation but fell sharply with increase of gas flow.

The effect of bed viscosity on a rising bubble may be estimated by calculating a bubble Reynolds number $U_b D_e/\nu$, where ν is the kinematic viscosity of the fully fluidised bed, as measured by the viscometers. For example, Schügerl et al. give $\nu \simeq 10\,\mathrm{cm^2/sec}$ for fully fluidised quartz particles in the size range 150–200 μ; these are very similar to the sand particles of Harrison and Leung (1962 a). In a typical case the bubble diameter $D_e = 10\,\mathrm{cm}$ and $U_b \simeq 70\,\mathrm{cm/sec}$, so that the bubble Reynolds number is 70. Since the Reynolds number gives a measure of the ratio of inertia forces to viscous forces, this value of 70 suggests that the inertia forces still predominate in a fluidised bed although its viscosity is so much greater than that of water. But the viscous forces may be responsible for the fact that bubbles in a fluidised bed appear to have a much larger included angle (Plate II) than would a similar bubble in water. For the case shown in Plate II, Schügerl *et al.*'s measured viscosity for a fully fluidised bed of 400–600 μ glass spheres is about 12 P, and taking the bed density as 1 g/ml, $U_b D_e/\nu = 50 \times \frac{5}{12} = 21$, a smaller bubble Reynolds number than in the work of Harrison and Leung (1962 *a*); the difference between the two Reynolds numbers may account for the apparent difference in bubble shape between the two sets of experiments.

2.4. The slugging behaviour of fluidised beds

For a gas fluidised bed of 1 or 2 in diameter, the bubble size becomes equal to the bed diameter when the height is more than 1 or 2 ft, and then the so-called 'slug flow' régime is entered. An extreme case of this kind was studied in detail by Lanneau (1960), whose air-fluidised catalyst bed was 3 in diameter and 15 ft high. He made direct measurements of the absolute velocity of individual bubbles by inserting capacitance probes in the bed, and from film records of the observed capacitance it was possible to measure the time taken for an individual bubble to traverse the known vertical distance from one probe to the next. These bubbles were continuously generated, in that the bed was continuously fluidised by a known velocity U, and the theory of Nicklin *et al.* (1962), for an air–water system, can therefore be applied.

Fig. 16 shows Lanneau's experimental observations of the absolute bubble velocity U_A as a function of the fluidising gas velocity, and the theoretical line $U_A = 1 \cdot 2(U - U_0) + 0 \cdot 35(gD)^{\frac{1}{2}}$ is obtained by combining (2.2) and (2.11), eliminating U_b and replacing G/A by $(U - U_0)$, the latter being assumed to represent the flow due to bubbles. The good agreement with experimental data is further confirmation of the validity of applying the two-phase model to a fluidised bed. It should be noted that in Lanneau's experiments the incipient fluidising velocity was so small ($U_0 \simeq 0 \cdot 02$ ft/sec) as to be almost negligible by comparison with U.

The agreement between theory and experiment in fig. 16 leads to two important conclusions.

(*a*) Bubbles which fill the tube appear to rise exactly as if they were in a true liquid, confirming that the incipiently fluidised system behaves like a liquid.

(*b*) If a large part of the air flow were percolating through the particles and not appearing in the bubbles, theory and experiment in fig. 16 would not be together. This agreement thus provides

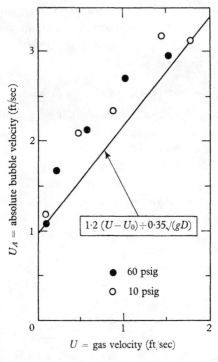

Fig. 16. Experimental results (Lanneau, 1960) for a 3 in diameter fluidised bed 15 ft high, with theory, for a gas-fluidised system.

further evidence that *in an aggregatively fluidised bed, the bubble flow accounts for all the fluid flow except what is required to incipiently fluidise the particles.*

2.5. The coalescence of bubbles in fluidised beds

It is well known (Yasui and Johanson, 1958; Baumgarten and Pigford, 1960) that bubbles grow in size as they rise in a gas-fluidised bed, and it is apparent that there are at least three possible explanations for this, namely,

(i) the effective hydrostatic pressure acting on the bubbles decreases as they rise up the bed;

(ii) bubbles may coalesce in a vertical line, i.e. one bubble may catch up another; and

(iii) neighbouring bubbles in a similar horizontal plane may combine when they are very close to each other.

An experimental study (Harrison and Leung, 1962 b) has been made of bubble growth by mechanism (ii). Nothing of a quantitative nature is yet known of (iii), but for many systems the effect of (i) is small. The experimental method for the study of bubble coalescence in a vertical line was suggested by the visual observation of the coalescence of spherical-cap air bubbles in water, for then the influence of one bubble on another is clear: when the following bubble approaches the leading bubble closely enough, it appears to be accelerated and gathered into the back of the leading bubble. Bubble coalescence thus occurs. This mode of coalescence may be explained by supposing that the wake behind a spherical-cap bubble travels with the bubble, and that coalescence takes place when one bubble moves into the wake of another.

(a) Experimental investigation of bubble coalescence in fluidised beds

(i) *Experimental.* Some preliminary experiments were carried out in a 'two-dimensional' bed of sand particles in which the paths of two air bubbles of equal size injected successively into the bed were followed by cine photography. The results suggested that a bubble has a wake behind it which can influence a following bubble —accelerating it, and eventually leading to coalescence.

The main experimental programme (Harrison and Leung, 1962 b) used a fluidised bed 2 ft square in cross-section containing sand, and into this bed a pair of air bubbles was injected in quick succession. The time interval between injections was varied until the second bubble of the pair coalesced with the first bubble at the moment the latter broke the top surface of the bed. Bubble coalescence was detected experimentally when the pair of injected bubbles was observed to break surface as one bubble, rather than as two. From the results with varying bed depths is was possible to make deductions about the size of the bubble wake and its influence on the following bubble. Throughout the experiment the bed of sand as a whole was fluidised by air maintained at a velocity some 5–10 % above that required for incipient fluidisation.

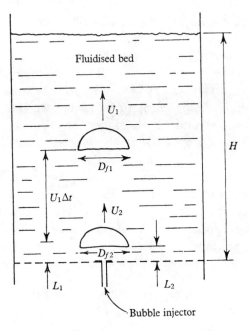

Fig. 17. Idealised position of a pair of bubbles immediately after the injection of the second bubble.

(ii) *Analysis of results.* For a given pair of bubbles injected into the bed the following quantities were measured:

(α) the overall bed height, H, and

(β) the time interval, Δt, between successive injections, so adjusted to give coalescence of the pair of bubbles at the top of the bed of height H.

The idealised condition in the bed immediately following the injection of the second bubble of a pair is shown in fig. 17. U_1 and U_2 refer, respectively, to the rising velocities of the leading and following bubbles. It is assumed, and the results of the 'two-dimensional' experiments support the assumption, that U_1 is un-

influenced by the presence of the following bubble. U_1 is given by
(2.19), and L_1 and L_2—the entrance effects for bubble injection—
are defined in the same way as L_0 in § 2.3 (c) (i) (p. 31). The entrance
effect for a given apparatus was found to be a function of the bubble
volume, and therefore for a pair of identical bubbles, $L_1 = L_2$. The
time t_c taken for the second bubble to catch up the first in order to
bring about coalescence at the bed surface is given by

$$t_c = [H - (U_1 \Delta t + L_1)]/U_1, \qquad (2.25)$$

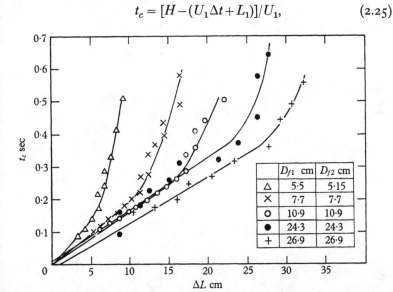

Fig. 18. Time for bubble coalescence as a function of the initial
distance of separation.

and the initial distance between the bubbles is

$$\Delta L = U_1 \Delta t + L_1 - L_2. \qquad (2.26)$$

Fig. 18 shows Harrison and Leung's results plotted from experimental measurements of the quantities on the right-hand sides of
(2.25) and (2.26), the results being those for which the wall effect
was small. The smooth curves drawn through the experimental
points show that the relative velocity between the bubbles is constant provided the distance between them is less than about one
bubble diameter. This is brought out more clearly in fig. 19 which
shows how the relative velocity $U_R = \mathrm{d}(\Delta L)/\mathrm{d}t_c$ is related to the

distance ΔL between the bubbles. At any particular ΔL, U_R is obtained from the reciprocal slope of the smooth curve in fig. 18. Although there is considerable scatter between the curves in fig. 19, they can be summarized by the single curve shown, which was derived by Harrison and Leung (1962 b) using a slightly different method of plotting the results.

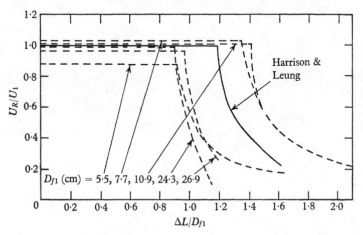

Fig. 19. Relative velocity between bubbles as a function of the distance of separation.

Carrying the analogy with a liquid further, Harrison and Leung assumed that the wake of a bubble has a velocity U_W and that the following bubble, rising in this wake, has an absolute velocity $U_W + U_2$ and therefore the relative velocity between the bubbles is

$$U_R = U_2 + U_W - U_1. \qquad (2.27)$$

Table 4 shows the values of U_W calculated by Harrison and Leung from (2.27) for bubble pairs close enough together to have a constant relative velocity, that is, the values of ΔL were within the straight-line part of the curves shown in fig. 18. Table 4 shows that under these conditions the wake velocity is essentially equal to the bubble velocity for bubbles less than 27 cm frontal diameter. The larger bubbles—whose frontal diameter is greater than half the size of the bed—are presumably subject to a wall effect and for that reason were not included in the results plotted in fig. 18.

Table 4. *Bubble coalescence in a 2 ft square bed*

Injected volumes (ml)		Observed bubble frontal diameter (cm)		Bubble velocity (cm/s)		Relative velocity U_R (cm/s)	Wake velocity U_W (cm/s)
		Leading bubble	Following bubble	Leading bubble	Following bubble		
1st	2nd	D_{f1}	D_{f2}	U_1	U_2		
53·0	53·0	5·5	5·5	41·2	41·2	34	34
82·0	82·0	7·7	7·7	50·8	50·8	51	51
194	194	10·9	10·9	60·8	60·8	61·5	61·5
1050	1050	18·2	18·2	61·8	61·8	60	60
2040	2040	24·3	24·3	61·9	61·9	62	62
2850	2850	26·9	26·9	72·0	72·0	69·5	69·5
2850	5640	26·9	34·9	72·0	84·7	83·5	70·8
4700	4700	33·4	33·4	83·3	83·3	67	67
4700	7050	33·4	38·8	83·3	87·7	72·5	68·1
5640	5640	34·9	34·9	84·7	84·7	51	51
7050	7050	38·8	38·8	87·7	87·7	48	48
7050	4700	38·8	33·4	87·7	83·3	57	61·4
7050	2850	38·8	26·9	87·7	72·0	45·5	61·2
7050	2040	38·8	24·3	87·7	61·9	35·5	61·3

(b) *A 'two-phase' hypothesis for bubble coalescence in
fluidised beds*

The main deductions from the above results on bubble coalescence in fluidised beds are as follows:

(i) The experimental results suggest that a velocity can be associated with the wake of a bubble, and this velocity is sensibly constant over a distance of about a bubble diameter behind the bubble; beyond that the wake velocity falls off sharply (fig. 19). This result is consistent with the observation of Davies and Taylor that a spherical-cap bubble is followed by a wake contained within the sphere of radius R indicated in fig. 8.

(ii) The average velocity of the main part of the wake is approximately equal to that of the bubble; experiments on very large bubbles (Harrison and Leung, 1962b) show that this result holds good provided the bubble diameter is less than about half the diameter of the apparatus.

(iii) The foregoing conclusions lead to the hypothesis that bubble coalescence in a vertical line takes place as a result of one bubble moving into the wake of another. The precise mechanism must however be complicated, because a following bubble may well be elongated, and so lose its spherical-cap shape, as a consequence of its front taking up the wake velocity of a leading bubble before the rear. It may be that bubble coalescence is rather easier in fluidised beds due to the absence of surface tension forces.

(iv) It is noteworthy that the extent and velocity of the wake of a bubble may provide a partial explanation of the good longitudinal solid mixing to be found in a gas–solids bed, because each bubble carries solid particles in its wake up the bed.

(v) Finally, it is necessary to stress that as yet the 'two-phase' hypothesis has only considered coalescence between two bubbles in a vertical line. In fact, the main source of uncertainty giving rise to some scatter in the experimental results is most probably due to the influence on the larger injected bubbles of the small air bubbles indigenous to the air–sand bed.

(c) *Bubble size as a function of bed height*

The general prediction of bubble size as a function of bed height is one of the ultimate aims of any investigation of bubble coalescence in a fluidised bed. This has been attempted (Harrison and Leung, 1962b; Leung, 1961) for a very special case, viz. for the size of bubble at various heights above an injection orifice in a fluidised bed. The data needed for this calculation are (α) the volume of the bubble at an orifice, (β) the rising velocity of a bubble and (γ) the rate of bubble coalescence. Information is available in this chapter on (β) and (γ), and (α) is considered in Chapter 3.

(i) *The calculation of bubble size.* The average starting size of a bubble at an orifice is known, provided the flow-rate through the orifice is such as to give bubbling at a known frequency (§ 3.4, p. 57): then with this information, the growth of a bubble up the bed by coalescence can be determined by step-wise calculation provided it is assumed that,

(α) coalescence takes place in a vertical line;

(β) the rising velocity of the bubble is always given by (2.19), with the appropriate value for V; and

(γ) some allowance is made in the case of three bubbles in line for the influence of the second bubble on the third. In this case the

wake of the first bubble influences (and probably elongates) the
second which, in its turn, has a wake which influences the third
bubble.

(ii) *Comparison with experiment* (Harrison and Leung, 1962b).
The maximum size of bubbles at various levels above an injection
orifice in a bed of sand fluidised by air was determined by ciné
photography of the bed surface. The bubbles were formed by
passing an independent stream of air through the injection tube.

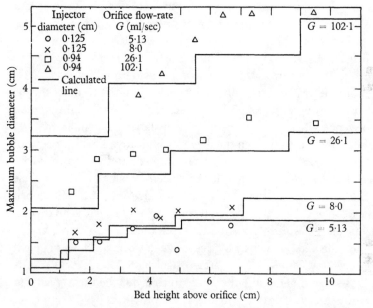

Fig. 20. Maximum bubble diameter as a function of bed height above
orifice in a 6 in diameter bed (Harrison and Leung, 1962b).

The results of these experiments using orifices of 0·125 and 0·94 cm
diameter in a 6 in diameter bed are given by the point values in
fig. 20. These are in tolerable agreement with the step-wise calculation shown by the continuous lines in the same figure.

This calculation is a very special case of a general relation between
bubble size and bed height. However, a general treatment awaits,
for instance, the study of bubble formation at a multihole distributor, and also the manner in which bubbles in close proximity in
the same horizontal plane affect one another.

4

CHAPTER 3

THE FORMATION OF BUBBLES IN FLUIDISED BEDS

3.1. The analogy with the gas–liquid system

In Chapter 2 experiments were described which showed that bubbles in a fluidised bed rise as if they were in an ordinary liquid of small viscosity. The question then arises as to how these bubbles form in the first place, and this chapter will describe experiments which suggest strongly that bubbles are formed in a fluidised bed in much the same way as they are in an inviscid liquid.

A simple way to form air bubbles in a liquid is to blow the air steadily through an orifice immersed in the liquid, and common experience shows that in these circumstances a regular train of bubbles is produced when the flow rate is steady. The behaviour of such a system has been extensively studied, and it was therefore logical to experiment with a similar system in which a stream of air was blown steadily through a single orifice into a fluidised bed (Harrison and Leung, 1961). In the same way as with the experiments on the steady rise of single bubbles, it was necessary to fluidise the bed incipiently by a separate uniform air flow, to give it the necessary liquid-like properties. In order to interpret the results of these experiments it is necessary to consider the theoretical and experimental background to the work provided by the study of bubble formation in an ordinary liquid.

3.2. The theory of bubble formation in an inviscid liquid

When air or other gas is blown steadily through an orifice into a liquid of small viscosity, a more or less regular train of bubbles is formed. At very low air rates, the frequency and size of the bubbles is governed mainly by a balance between surface tension and buoyancy forces. At higher air rates the inertia of the liquid moved by the rising bubbles becomes more important than surface tension: it is with this régime that we are mainly concerned because of course surface tension is zero in a fluidised bed. At still higher air rates, the momentum of the air issuing from the orifice is sufficient to maintain it as a jet before break-up into separate bubbles.

With true liquids, the middle régime in which surface tension and the momentum of the issuing air are negligible is important because it covers the flow-rates—up to a few hundred millilitres per second—used in the orifices of many pieces of chemical engineering equipment in which bubbles are formed in a liquid, for example, bubble plates and sieve trays. Moreover, this middle régime of orifice air flow has the advantage of being easy to analyse theoretically, as will now be shown.

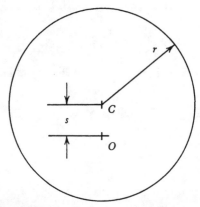

Fig. 21. Bubble forming in a liquid; the gas source is at O.

Fig. 21 shows the bubble during its formation but prior to detachment from the orifice O. Since the momentum of the air leaving the orifice O is negligible, the liquid should, to begin with at any rate, be pushed out equally in all directions, and the bubble is therefore assumed to be spherical. The bubble is initially centred on O at time $t = 0$, but because of buoyancy it tends to rise, and at the instant shown in fig. 21 the centre C has risen a distance s. In the case which is simplest to analyse, the gas flow-rate G from the orifice is constant. This can easily be arranged experimentally by having a narrow constriction between the air reservoir and the orifice O, so that the flow from the orifice is independent of the pressure within the bubble. In this case the bubble volume V at the time t is

$$V = 4\pi r^3/3 = Gt, \qquad (3.1)$$

which gives the relation between the radius of the bubble r, and t.

It is assumed that the bubble will detach when its base reaches the orifice O, that is, when $r = s$, and s can be calculated from the mechanics of the upward motion of the bubble. This motion is defined by balancing the buoyancy force $\rho_L V g$ against the rate of change of upward momentum of the liquid surrounding the bubble, the inertia of the air within the bubble being neglected. Here ρ_L is the liquid density. It is shown in Appendix A 1 (g) (p. 128) that when a sphere moves in an inviscid fluid, with no separation of the flow, the effective mass added to the sphere by the surrounding fluid is half the displaced mass. For the forming bubble the upward momentum at any instant is therefore $\frac{1}{2}\rho_L V \, ds/dt$, and the equation of upward motion is

$$\rho_L V g = \frac{d}{dt}\left(\tfrac{1}{2}\rho_L V \frac{ds}{dt}\right). \tag{3.2}$$

By eliminating V from between (3.1) and (3.2) and integrating with respect to t, then

$$\frac{ds}{dt} = gt, \tag{3.3}$$

using the initial condition that $ds/dt = 0$ when $t = 0$. Integrating (3.3) gives

$$s = \tfrac{1}{2}gt^2, \tag{3.4}$$

using the second initial condition that $s = 0$ when $t = 0$. Equation (3.4) shows that the bubble has an upward acceleration equal to that of gravity; this is perhaps surprising in view of the fact that the effective mass of the bubble is half its displaced mass; for a bubble of constant volume the upward acceleration should therefore be $2g$. However, the variation in bubble volume with time leads to an additional momentum term in (3.2), namely

$$\tfrac{1}{2}\rho_L \frac{dV}{dt}\frac{ds}{dt},$$

and this reduces the acceleration from $2g$, for a constant volume bubble, to g for a bubble whose volume increases linearly with time.

Using the assumption, mentioned previously, that the bubble detaches from the orifice when $r = s$, we can, from (3.1) and (3.4), calculate the time of formation of the bubble,

$$t = \frac{1}{g^{\frac{3}{5}}}\left(\frac{6G}{\pi}\right)^{\frac{1}{5}},$$

and substituting this value of t into (3.1) gives the bubble volume at detachment

$$V = \left(\frac{6}{\pi}\right)^{\frac{1}{5}}\frac{G^{\frac{6}{5}}}{g^{\frac{3}{5}}} = 1\cdot138\frac{G^{\frac{6}{5}}}{g^{\frac{3}{5}}}. \tag{3.5}$$

3.3. Experimental results on bubble formation with a gas–water system

Fig. 22 shows a comparison between the above theory and the experimental results of various investigators who formed bubbles

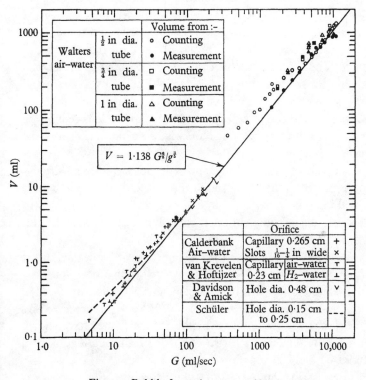

Fig. 22. Bubble formation at an orifice in water.

steadily at a single orifice by blowing air or hydrogen into water. The agreement is reasonably satisfactory and confirms that the theory gives a good approximate picture of the basic mechanism involved in the formation of bubbles at an orifice. The factors likely to vitiate the comparison between theory and experiment can be listed as follows.

(a) In the theoretical derivation it was assumed that each bubble is spherical during formation. A bubble forming in a liquid of zero surface tension must in fact distort as it accelerates upwards, for the same sort of reason that a steadily rising bubble takes on the spherical-cap shape described in Chapter 2. This distortion is brought about by the fact that the pressure on the surface of the bubble is constant—because it is filled with gas—and with the unseparated flow which necessarily occurs during the initial motion, the elements of liquid round the lower surface of the bubble accelerate faster than those at the upper surface. This causes the bubble to distort into the cross-section of a mushroom, a tongue of liquid being projected upwards from the base of the bubble towards its centre. These distortions have been studied both experimentally and theoretically for a 'two-dimensional' bubble by Walters and Davidson (1962). An analysis of the three-dimensional bubble by Walters and Davidson (1963) shows that the distortions reduce the constant 1·138 in (3.5) without otherwise altering the form of the equation.

(b) The train of rising bubbles above the orifice induces a circulation current within the liquid round the orifice, so that each bubble is formed in an upwards current, and not in stagnant liquid as was supposed in the derivation of the theory. The rising bubbles induce the upward current in two ways.

(i) When a sphere moves through an inviscid fluid, the fluid at a fixed point in space does not return to its original position when the sphere has moved past, but each element of fluid is displaced in the direction of the sphere's motion. This 'drift' is described by Tietjens (1934, p. 106); Rowe and Partridge (1962) have shown experimentally that such a 'drift' of particles is brought about in a gas-fluidised bed by rising bubbles.

(ii) Soon after each bubble detaches from the orifice, a wake is formed below the bubble, and the bubble shape takes on the spherical-cap form described in Chapter 2. The formation of the wake means that liquid is carried up with each bubble, and this liquid contributes further to the circulating current near the orifice.

It is not easy to estimate the effect of the circulating current, but the results in fig. 22 suggest that it does not have much influence on the bubble volume.

(c) A complication that arises, particularly at the higher flow-rates, is the coalescence of bubbles shortly after formation, leading to the so-called double or quadruple bubble formation reported by Davidson and Schüler (1960). This occurs because a bubble gets into the wake of the one above, and is the first step towards the formation of a continuous gas jet observed at very high flow-rates. Davidson and Schüler (1960) observed that in double- or quadruple-bubble formation, the bubbles were nearly regular and of the same size, but coalesced shortly after leaving the orifice. The dotted line shown in fig. 22 is a plot of Schüler's (1959) photographic measurements of the volume of the upper bubble of each group before coalescence.

Walters (1962) used much higher flow rates and also observed coalescence of bubbles near the orifice, but the second bubble of each pair was much smaller than the first, and he therefore measured the volume of the bubble that moved up after coalescence of the unequal pair. These volume measurements were made by

(i) counting on a high-speed film the number of complete bubbles in a measured time interval, and

(ii) by measuring bubble dimensions from the film.

It might be expected that the bubble volume thus measured would be greater than the theoretical, and fig. 22 shows that this is in fact the case, since most of Walters's experimental points lie above the theoretical line. Walters has given a quantitative explanation of double-bubble formation by assuming that as each bubble detaches, it leaves behind a hemispherical bubble of the same radius as the orifice. This residual bubble then grows, rapidly at first, and for a given orifice diameter there is a certain flow at which the top of the residual bubble overtakes the base of the newly detached bubble, leading to double-bubble formation. In practice Walters found that his theory gives only an approximate idea of the flow-rate at which double-bubble formation begins, but the explanation does show why a series of orifices is needed to study bubble formation over a wide range of air rates. For each orifice, the upper limit of flow is governed by double-bubble formation; the lower limit is determined by the requirement that the bubbles should be somewhat larger than the orifice. Double-bubble formation is important in comparing the results for air–water with those of Harrison and

Leung (1961) for a fluidised bed in which the technique was such that *all* bubbles forming at the orifice were counted by a capacitance meter; the comparison will be discussed in the next few pages.

(*d*) Another factor that makes the above theory a little unrealistic is the presence of the orifice itself. Some of the points plotted in fig. 22 were obtained from experiments with an orifice consisting of a hole in a horizontal plate with liquid above and air supplied from below, the hole forming the orifice (Davidson and Amick, 1956; Davidson and Schüler, 1960). The presence of the plate affected the motion of the liquid round the forming bubble; Davidson and Schüler (1960) therefore used an effective mass of $\frac{11}{16}\rho_L V$ rather than $\frac{1}{2}\rho_L V$ as in the above analysis. The factor $11/16$ is obtained theoretically for a sphere instantaneously in contact with a fixed wall and moving normal to it in an inviscid liquid (Milne-Thomson, 1960, p. 504); the use of this factor $11/16$ led to a factor of $1\cdot378$ instead of $1\cdot138$ in (3.5).

(*e*) Finally, there is the question of the variation of air flow into the bubble during formation. Unless special precautions are taken to ensure a large pressure drop through the orifice supplying the bubble, the flow will vary during formation. Such variations in flow are likely to have occurred in some of the experiments from which the points in fig. 22 were plotted. And for this reason, some of the literature data on bubble formation at an orifice are very specific to the apparatus used. For example, Quigley, Johnson and Harris (1955) used an orifice above a gas reservoir of finite size. This reservoir was large enough to permit considerable flow variations during bubble formation at low gas rates, and consequently the bubbles were much larger than is predicted by (3.5); but there is good agreement with (3.5) at higher flow-rates, when the pressure drop through the orifice below the bubble was enough to ensure constant flow throughout formation.

In spite of all these complicating factors, it can be concluded from fig. 22 that (3.5) is a reasonably accurate summary of the experimental results, and that the theory leading to (3.5) gives a good picture of the most important factors governing bubble formation in an inviscid liquid.

3.4. Experimental results on bubble formation at a single orifice in a fluidised bed

Harrison and Leung (1961) reported experimental results on bubble formation in an incipiently fluidised bed. Their apparatus, shown diagrammatically in fig. 23, consisted of a bed of particles incipiently fluidised by a uniformly distributed air-flow, the bubbles being formed at an orifice immersed in the fluidised bed and fed

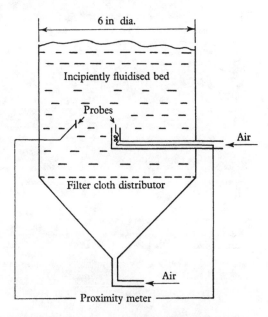

Fig. 23. Apparatus for studying bubble formation at an orifice in an incipiently fluidised bed (Harrison and Leung, 1961).

with its own air supply. This supply is equivalent to the air supply in the air–water experiment, but the frequency of bubble formation in the fluidised bed was measured by recording the capacity between two probes, one placed within the orifice, and the other outside it. The method was first checked on an air–water system. Fig. 24 shows a typical set of oscilloscope traces, one for an air–water system and three for fluidised systems, each spike on the trace recording the detachment of a bubble; the frequency of bubble formation was obtained by reference to the 50-cycle time-bases shown in Fig. 24.

For the fluidised system, a series of orifices was used, and it was found that each orifice was suitable over a certain range of flow-rates (fig. 25). For each orifice, the maximum flow-rate was governed by irregular bubble formation (as shown by trace (d) in fig. 24), though it was very hard to differentiate between the regular formation of

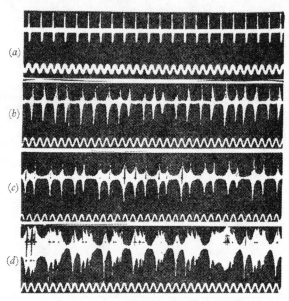

Fig. 24. Capacitance traces for the formation of air bubbles in water and in various air-fluidised beds (Harrison and Leung, 1961).

System	Air flow-rate through 0·125 cm orifice (ml/s)	Frequency of bubble formation (s⁻¹)	Average bubble volume (ml)
(a) Air–water	5·55	31	0·18
(b) Air–catalyst	2·51	32	0·08
(c) Air–sand	8·16	30	0·27
(d) Air–catalyst	18·4	About 29	(0·63)

equal bubbles and the start of multiple bubble formation. In this connexion it is significant that the results of Harrison and Leung (fig. 25) lie below the theoretical line at higher flow-rates, whereas Walters's results (fig. 22) lie mainly above the theoretical line. The difference is probably due to undetected double-bubble formation in the experiments of Harrison and Leung, since a small secondary bubble—not counted by Walters's technique—might make a spike on the oscilloscope trace, thus raising the apparent frequency and lowering the calculated bubble volume.

However, the results of Harrison and Leung do show good general agreement with (3.5), and it is noteworthy that the bubble

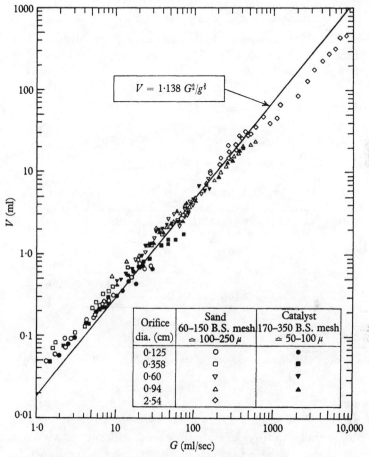

Fig. 25. Bubble formation at an orifice in an incipiently fluidised bed. Results of Harrison and Leung (1961).

frequency was not affected by bed depth or by altering the main fluidising air velocity, though it was found essential to have the bed incipiently fluidised.

Bloore and Botterill (1961) reported similar results on bubble formation at an orifice in an incipiently fluidised bed of much larger particles, and their results, obtained by a γ-ray technique, are shown in fig. 26 together with a theoretical curve calculated from (3.5) by assuming that the frequency of bubble formation is G/V.

Inferences to be drawn from the results

(i) It is clear from figs. 25 and 26 that (3.5) is in reasonable agreement with the experimental results on bubble formation in a fluidised bed. It seems probable, therefore, that the mechanism of

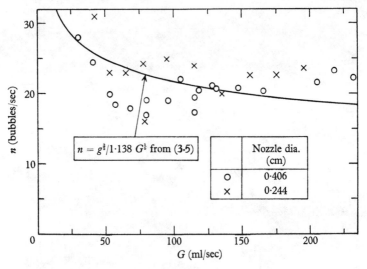

Fig. 26. Bubble formation at an orifice in an incipiently fluidised bed of large particles (0.8 mm diameter Ballotini). Results of Bloore and Botterill (1961).

bubble formation is the same as in an inviscid liquid, namely, by growth of a nearly spherical bubble at the tip of the orifice, the bubble simultaneously moving upwards with an acceleration governed by a balance between buoyancy and the inertia of the particulate phase surrounding the bubble.

(ii) The method of plotting the results in figs. 25 and 26 is based on the assumption that the bubbles leaving the orifice represent the same volume flow rate as the air fed to the orifice; this assumption implies that there is no net leakage of air to the fluidised bed from each bubble. Harrison and Leung (1961) have estimated the likely order of magnitude of the leakage of air from the roof of a forming bubble into the surrounding bed, and although this leakage was in most cases a small proportion of the bubble volume, in one case 15 % of the volume of a bubble was estimated to leak through the roof during formation. The corresponding percentage would be much higher for the larger particles (0·8 mm) used by Bloore and Botterill (1961). However, if air leaks out of the upper half of a forming bubble, there are equally good arguments for supposing that there will be a very similar influx of air over the lower half of the bubble, and the net effect on its volume may be negligible. The good agreement (fig. 26) between (3.5) and the results of Bloore and Botterill (1961) supports these arguments, for their data would be sensitive to the leakage effects because of the relatively high incipient fluidising velocity with the 0·8 mm Ballotini used. Furthermore, an analysis presented in Chapter 4 shows that although there is a continuous exchange of fluidising fluid between a steadily rising bubble and the surrounding particulate phase, the net leakage is zero; although this analysis does not directly apply to a forming bubble, it seems possible that the same conclusion will hold good.

The above inference, that the volume of the bubbles formed is equivalent to the volume of air fed to the orifice, is important, because it implies that *all the excess air above the minimum required to fluidise the bed passes through as bubbles*, at least for a single orifice.

As a final qualification it ought to be stressed that *bubble* formation at an orifice has been under consideration in this chapter; for it is clear that there must be a moment, as the air flow through an orifice is increased, when the flow is sufficient to support a jet of air within the bed, and then the conditions approximate to those in a spouted bed.

3.5. Bubble formation in a fluidised bed with distributed gas supply

In industrial practice, the fluidising gas passes through some sort of distributor resembling a sieve tray or bubble-cap tray, which also supports the particles. With such arrangements the gas can be regarded as coming through a number of parallel orifices and the bubble formation at each orifice should resemble the bubble formation at the single orifice described in the preceding sections. The problem of bubbling from parallel orifices close together is difficult to treat analytically, and no experimental information is available for gas–liquid systems with regard to bubble frequency and volume. However, there seems good reason to suppose that a fluidised system fed by multiple orifices in parallel should be similar to a single orifice system; and that therefore all the excess air above the minimum required to fluidise the bed should pass through as bubbles.

Since the incipiently fluidised bed behaves so much like a liquid, it also seems reasonable to suppose that the rate of leakage of particles through the holes of the distributor should occur in the same way as for liquids on sieve trays; this liquid leakage is the phenomenon of 'dumping'. However, since the mechanism of dumping of liquids from sieve trays is somewhat uncertain, it is probable that leakage of fluidised particles through support grids will continue to remain an empirical study for some time.

In the laboratory it is common practice to support fluidised particles on a gauze which also acts as a gas distributor, so that the gas is fed uniformly. Here again, little information is available for gas–liquid systems. But Rice and Wilhelm (1958) have carried out a theoretical analysis of the instability of the lower horizontal surface of a fluidised bed, by assuming that the instability is of the same kind as would occur in an inverted cup filled to the brim with liquid. Rice and Wilhelm showed that waves on the lower horizontal surface of a fluidised bed are always unstable, there being a certain wavelength which has maximum instability. This wavelength is of the order of a few millimetres, which is perhaps the size of the bubbles formed at the base of the fluidised bed with a uniform air supply. Thus Yasui and Johanson (1958), who measured bubble sizes optically in beds of particles fluidised by a uniformly distributed air supply, reported bubble sizes of a few millimetres near the distributor, though the bubbles were much larger higher up, presumably due in part to coalescence in the manner described in § 2.5 (c) (p. 48).

CHAPTER 4

THE EXCHANGE BETWEEN THE BUBBLE AND PARTICULATE PHASES

4.1. Introduction

In the preceding chapters, experiments were described which showed that bubbles of fluidising fluid behave in the fluidised bed in many ways as if they were in an ordinary liquid. The question then arises as to how the fluidised particles can behave like a liquid in the neighbourhood of a bubble, and in particular:

(a) what prevents the roof of the bubble falling in when there is no surface tension; and

(b) how do the fluidised particles manage to behave as an incompressible fluid of small viscosity; and

(c) what quantity within a system of fluidised particles is equivalent to the pressure within an ordinary liquid?

The following analysis, with supporting experiments, gives a partial answer to these questions; and it also gives the magnitude of the exchange of fluidising fluid between a rising bubble and the surrounding particulate phase. This exchange was discussed in Chapter 3 in connexion with bubble formation. The exchange is also important in considering a chemical reaction between the fluidising fluid and the particles, because the fluid within a bubble is unable to react with the particles; this question will be considered in Chapter 6.

Jackson (1963) has recently given a more elaborate analysis of the flow near a bubble rising in a fluidised bed; his analysis does not conflict with the analysis given below, and confirms the more important results.

4.2. The relative motion between fluid and particles in the particulate phase

A fluidised bed in which the behaviour is essentially two-dimensional can be arranged by containing the bed of particles between two vertical plates, a centimetre or two apart, the width and height of the bed, measured in the plane of the plates, being much greater.

Experiments with such apparatus have been reported by Wace and Burnett (1961), who showed that bubbles extend right through the bed from plate to plate, the cross-section of each bubble being the same in any vertical plane parallel to the plates. The bubbles can thus be regarded as two-dimensional, and Plate III shows a typical photograph of a bubble seen through the transparent walls of the fluidised bed.

The two-dimensional motion also has the merit that it is easier to treat theoretically than the corresponding three-dimensional problem, although the equations for the latter case are easily developed along the same lines as in the two-dimensional problem. The theory is based on the following assumptions.

(a) The particulate phase is treated as an incompressible fluid having the same bulk density as the whole bed at incipient fluidisation. This is consistent with the premise that all the excess fluid, above what is required for incipient fluidisation, passes through as bubbles. The continuity equation for the particles is therefore

$$\frac{\partial v_x}{\partial x} + \frac{\partial v_y}{\partial y} = 0, \qquad (4.1)$$

where x and y are rectangular coordinates, the axes being in a plane parallel to the containing walls with the y axis vertical, and v_x and v_y are the components of the particle velocity parallel to x and y. Equation (4.1) is derived from a material balance on the particles entering and leaving a fixed element $dx\,dy$.

(b) The *relative* velocity between the fluidising fluid and the particles is assumed to be proportional to the pressure gradient within the fluidising fluid, and the *absolute* components of the fluid velocity are therefore

$$\left.\begin{aligned}
u_x &= v_x - K\frac{\partial p_f}{\partial x}, \\[2mm]
u_y &= v_y - K\frac{\partial p_f}{\partial y},
\end{aligned}\right\} \qquad (4.2)$$

where K is a permeability constant characteristic of the particles and of the fluidising fluid, within which the pressure is p_f. The assumed proportionality between the relative velocity and the pressure gradient is the same as D'Arcy's law, which is a well-established relation for percolation through fixed beds of fine sand and filters; see, for example, Jaeger (1956, p. 395). The same law is also represented by the left-hand part of the graph in fig. 4; this diagram shows clearly that D'Arcy's law holds good only for low Reynolds numbers. The same limitation—that Re′ must be small—therefore applies to the analysis of this chapter.

The factor K is of course dependent upon the particle spacing, but it can be taken as constant in view of assumption (a) above that the particles are evenly spaced throughout the particulate phase. K is chosen so that u_x and u_y are interstitial (absolute) velocities through the pores; the volume flow across unit area normal to the x axis is therefore $u_x \epsilon_0$, and similarly among the y axis the flow per unit area is $u_y \epsilon_0$.

(c) The continuity equation for the fluidising fluid, assumed to be incompressible, is

$$\frac{\partial u_x}{\partial x} + \frac{\partial u_y}{\partial y} = 0, \tag{4.3}$$

since the voidage is everywhere ϵ_0.

The velocities can then be eliminated from between (4.1), (4.2) and (4.3), giving

$$\frac{\partial^2 p_f}{\partial x^2} + \frac{\partial^2 p_f}{\partial y^2} = 0, \tag{4.4}$$

a rather surprising result, because it shows that with given boundary conditions, the pressure distribution is unaffected by the motion of the particles, since (4.4) is exactly the same as it would be for a fixed bed of particles (Appendix A and Jaeger, 1956, p. 433). A similar result holds good for three-dimensional motion (Davidson, 1961) and the equation governing the pressure is

$$\operatorname{div} \operatorname{grad} p_f = 0, \tag{4.5}$$

the steps in the derivation being very similar to those given above.

In using these equations to calculate what happens in the neighbourhood of a rising bubble in an incipiently fluidised bed, the first step is to calculate the motion of the particles round the rising bubble, assuming that the particulate phase behaves as an inviscid liquid. Then the pressure distribution is calculated from (4.4) or (4.5), using as boundary conditions: (i) the known pressures far above and below the rising bubble, and (ii) the fact that the pressure throughout the bubble is constant. The resulting particle velocities and fluid pressures are now inserted in (4.2), or its three-dimensional equivalent, to give the fluid velocities, which complete the solution of the problem.

4.3. Motion of the particles, and of the fluid, around a rising bubble

(a) Particle motion

The first step in the solution outlined above is to solve the problem of a bubble rising in an inviscid liquid, this liquid being the fluidised particles. This problem presents great (unsolved) analytical difficulties—discussed in Chapter 2—in that the shape of the bubble (the spherical-cap form) has to be found as part of the answer. Therefore here the assumption is made that the bubble has a circular cross-section, so that the particle streamlines are given in the two-dimensional case by the velocity potential

$$\phi = - U_b\left(r + \frac{a^2}{r}\right)\cos\theta \qquad (4.6)$$

(cf. (A. 13)), where r and θ are the polar coordinates given in fig. 39, and a is the radius of the bubble. It should be noted that (4.6) refers to a bubble held stationary by a downward flow of the particulate phase, a situation which is easier to grasp than when the bubble is rising relative to the observer. In the present context, this result is deficient in that it gives a varying pressure round the surface of the bubble, instead of the constant pressure required by the conditions of the problem, as set out in Chapter 2. However, (4.6) gives particle streamlines which bear some similarity to those round an actual bubble, particularly near the upper surface of the bubble. A similar use of the ideal streamlines round a sphere was made by Davies and Taylor (1950) to explain the spherical-cap shape of a large air bubble rising in water.

At this stage it is worth noting that the pressure P necessary to bring about the particle motion defined by (4.6) can be calculated from Bernoulli's theorem in the usual way for an inviscid liquid. This pressure P is not just p_f, the pressure within the fluidising fluid, but includes the effect of inter-particle forces represented by a pressure p_p, that is

$$P = p_f + p_p. \qquad (4.7)$$

The relative magnitudes of p_f and p_p will be discussed in §4.5, p. 74.

Equation (4.7) is of course applicable to the three-dimensional case, and the result corresponding to (4.6) is

$$\phi = -U_b\left(r + \frac{a^3}{2r^2}\right)\cos\theta \qquad (4.8)$$

(cf. (A. 15)).

(b) Pressure distribution within the fluidising fluid

We have to find a pressure distribution to satisfy (4.4) or (4.5) throughout the region round the bubble, and to satisfy the boundary condition far above and below the bubble that the pressure gradient dp_f/dy shall have a constant value J. The value of J is determined by the need to incipiently fluidise the particles, and J is the pressure gradient in a vertical direction such that the weight of the particles shall be just supported. For the two- and three-dimensional cases, the pressure distributions used are those given by (A. 18) and (A. 21), namely,

$$p_f = -J\left(r - \frac{a^2}{r}\right)\cos\theta, \qquad (4.9)$$

$$p_f = -J\left(r - \frac{a^3}{r^2}\right)\cos\theta. \qquad (4.10)$$

These equations satisfy (4.4) and (4.5), and at a great distance above and below the bubble the pressure gradient is, as shown in Appendix A. 2, p. 129, $\partial p_f/\partial y = -J$. The interstitial fluid velocity u_0, relative to the particles, must be K times this value, so that

$$u_0 = KJ, \qquad (4.11)$$

and of course $u_0\epsilon_0 = U_0$, the superficial fluid velocity at incipient fluidisation. In this analysis u_0 will be used because the important parameter is the fluid velocity as seen by an observer looking at the fluid moving through the particle bed.

(c) The absolute velocities of the fluidising fluid

The fluidising fluid velocities are found from (4.2), though the equivalent equations in polar coordinates are more convenient, and are, in either the two- or three-dimensional axi-symmetric case,

$$\left. \begin{aligned} u_r &= v_r - K\frac{\partial p_f}{\partial r}, \\ u_\theta &= v_\theta - \frac{K}{r}\frac{\partial p_f}{\partial \theta}. \end{aligned} \right\} \tag{4.12}$$

The particle velocities $v_r = \partial\phi/\partial r$ and $v_\theta = \partial\phi/r\,\partial\theta$ are obtained from (4.6) or (4.8), and the fluid pressure p_f is obtained from (4.9) or (4.10), and substituting these quantities into (4.12), and using (4.11) to eliminate J, gives in the two-dimensional case,

$$\left. \begin{aligned} u_r &= \left[\frac{a^2}{r^2}(U_b + u_0) - (U_b - u_0)\right]\cos\theta, \\ u_\theta &= \left[\frac{a^2}{r^2}(U_b + u_0) + (U_b - u_0)\right]\sin\theta. \end{aligned} \right\} \tag{4.13}$$

The corresponding three-dimensional result is

$$\left. \begin{aligned} u_r &= \left[\frac{a^3}{r^3}(U_b + 2u_0) - (U_b - u_0)\right]\cos\theta, \\ u_\theta &= \left[\frac{a^3}{r^3}\left(\frac{U_b}{2} + u_0\right) + (U_b - u_0)\right]\sin\theta. \end{aligned} \right\} \tag{4.14}$$

To trace the fluid streamlines, we define a stream function ψ_f by the equations

$$\left. \begin{aligned} u_r &= -\frac{1}{r}\frac{\partial\psi_f}{\partial\theta}, \\ u_\theta &= \frac{\partial\psi_f}{\partial r}, \end{aligned} \right\} \tag{4.15}$$

in the two-dimensional case, and by the equations

$$\left. \begin{aligned} u_r &= \frac{-1}{r^2\sin\theta}\frac{\partial\psi_f}{\partial\theta}, \\ u_\theta &= \frac{1}{r\sin\theta}\frac{\partial\psi_f}{\partial r}, \end{aligned} \right\} \tag{4.16}$$

in the three-dimensional case. It should be observed that ψ_f, so defined, does not represent the actual flow of fluidising fluid, but the actual flow divided by ϵ_0.

A two-dimensional stream function consistent with (4.13) and (4.15) is

$$\psi_f = (U_b - u_0)\left(1 - \frac{A^2}{r^2}\right)r\sin\theta \quad \text{(i)},$$

where
$$\frac{A^2}{a^2} = \frac{U_b + u_0}{U_b - u_0} \quad \text{(ii)}.$$

$$\left.\right\} \quad (4.17)$$

The corresponding three-dimensional result from (4.14) and (4.16) is

$$\psi_f = (U_b - u_0)\left(1 - \frac{A^3}{r^3}\right)\frac{r^2\sin^2\theta}{2} \quad \text{(i)},$$

where
$$\frac{A^3}{a^3} = \frac{U_b + 2u_0}{U_b - u_0} \quad \text{(ii)}.$$

$$\left.\right\} \quad (4.18)$$

The physical significance of A will be discussed in §4.4.

(d) The exchange between the bubble and the particulate phase

At the periphery of the bubble, the radial velocity of the particles is zero; hence the absolute radial velocity of the fluid is the same as its velocity relative to the particles, $-K\,\partial p_f/\partial r$. But the pressure distribution round the moving void is known to be the same as round a fixed void in the same bed. Hence q, the flow of fluidising fluid into and out of a moving bubble, is the same as if the bubble were fixed, that is, from (A. 20)

$$q = 4u_0\epsilon_0 a = 4U_0 a, \tag{4.19}$$

for a two-dimensional bubble, and from (A. 23)

$$q = 3u_0\epsilon_0\pi a^2 = 3U_0\pi a^2, \tag{4.20}$$

for a three-dimensional bubble. The magnitude of q is important in analysing the behaviour of a fluidised catalytic reactor; this is considered in Chapter 6. Equation (4.20) is similar to the result suggested by Zenz (1957b) who conjectured that the flow through a rising bubble is $2U_0\pi a^2$; this result is also given by Zenz and Othmer (1960, p. 279).

4.4. Interpretation; particle and fluid streamlines; comparison with experiment

The results of the preceding section are easiest to interpret by imagining the bubble held stationary by a downward flow of the particulate phase, or by thinking of an observer moving with the rising bubble, and the results (4.6), (4.8), (4.17) and (4.18) are the

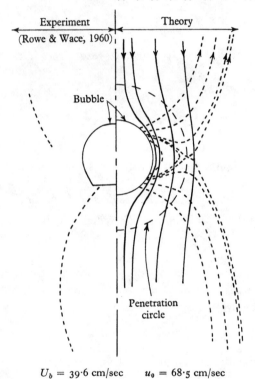

$$U_b = 39 \cdot 6 \text{ cm/sec} \qquad u_0 = 68 \cdot 5 \text{ cm/sec}$$

Fig. 27. Two-dimensional bubble held stationary by a downward flow of the particulate phase. The continuous curves are particle tracks (A. 14) and the broken curves are fluid streamlines (4.17).

particle velocity potential and fluid stream function seen by this observer. Typical streamlines are shown in figs. 27 and 28, the particle streamlines being continuous and the fluid streamlines broken.

The particle streamlines are those of irrotational motion; as mentioned previously these streamlines do not constitute a proper solution of the flow round a rising bubble, because this ought to have the spherical-cap form with a wake below as described in Chapter 2. Nevertheless, the particle streamlines shown in figs. 27 and 28 do represent a first approximation to the true streamlines *above* the bubble.

The fluid streamlines are crucially affected by the magnitude of the incipient fluidising velocity u_0, and in particular by whether u_0 is greater or less than U_b. Thus in the former case the fluidising fluid at infinity is moving upwards relative to the bubble, and in the latter case downwards. In the two cases the streamlines near the bubble are radically different, as can be seen from figs. 27 and 28.

$$(a)\ u_0 > U_b$$

Fig. 27 shows the particle and fluidising fluid streamlines for a case in which $u_0 > U_b$. The theoretical streamlines shown in the right-hand half of the diagram were calculated from (4.17), and the experimental streamlines and bubble shape shown in the left-hand half of the diagram were given by Rowe and Wace (1960). They observed bubbles rising in a fluidised bed of 0·46 mm Ballotini contained between two vertical Perspex plates, and photographed the traces shown by injecting nitrogen dioxide. Similar experiments were reported by Wace and Burnett (1961). They pointed out that the NO_2 trace observed is not necessarily a fluid streamline, as seen by an observer moving with the bubble, unless the injection point is far below the bubble. This effect was important in all the above-mentioned experiments, and it means that the NO_2 trace above the bubble is displaced to the left in fig. 27; this may partly account for the difference between the experimental trace on the left of fig. 27 and the corresponding theoretical fluid streamline.

The significance of the dotted circle in fig. 27, labelled 'penetration circle', may be understood by examining the form of (4.17) and comparing it with (A. 19). Since $u_0 > U_b$, A^2 is negative and (4.17(i)) has exactly the same form as (A. 19). Equation (4.17) therefore represents the percolation of fluid through a fixed bed of particles into a cylinder of radius $|A|$, and this is the radius of the 'penetration circle' shown in fig. 27. Between the 'penetration circle' and the bubble, the fluid streamlines are distorted by the downward motion of the particles, the fluid entering the lower half of the bubble being dragged down, by the particles, on its way in; the fluid leaving the top half of the bubble is similarly dragged down, by the particles, on its way out. On either side of the bubble there is a small region, within which, fluid leaving the bubble is dragged down by the particle motion, and re-enters the bubble without getting into the main stream.

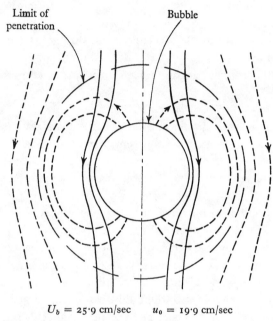

$U_b = 25{\cdot}9$ cm/sec $u_0 = 19{\cdot}9$ cm/sec

Fig. 28. Three-dimensional bubble held stationary by a downward flow of the particulate phase. The solid curves are particle tracks (A. 16) and the broken curves are fluid streamlines (4.18).

When $u_0 > U_b$, as with large particles, the flow of fluidising fluid in the neighbourhood of a rising bubble is thus not greatly different from its motion through a cavity in a fixed bed.

(b) $u_0 < U_b$

Fig. 28 shows that when $u_0 < U_b$—as with small particles, and this is representative of most practical applications of fluidisation—the fluid streamlines are completely different from the streamlines for a percolation problem. Instead, the fluid streamlines are identical with the streamlines of irrotational motion round a cylinder or sphere of radius A. For example (4.18 (i))—which represents the fluidising fluid streamlines round a rising spherical bubble in a fluidised bed—is of exactly the same form as (A. 16) which represents irrotational motion past a solid sphere. In fig. 28 the circle labelled 'limit of penetration' represents a sphere of radius A given by (4.18(ii)). Outside this sphere the fluidising fluid is dragged down by the motion of the particles needed to keep the bubble stationary. What is more interesting is the motion in the region between the 'limit of penetration' and the bubble itself. In this region, fig. 28 shows that the fluidising fluid leaves the roof of the bubble and, after making a circuit within the particles, returns again to the bubble; the 'limit of penetration' is as far as any fluid within the bubble can travel. The physical reason for this somewhat surprising result is that the pressure gradient in an upward direction, necessary to maintain the bed in a fluidised condition, forces the fluid out of the top of the bubble, whereupon the downward moving particles drag the fluidising fluid to the bottom of the bubble which it re-enters under the influence of the pressure gradient. We therefore have the important result that the fluid within a rising bubble stays with the bubble, but makes regular excursions into the surrounding particulate phase without, however, venturing beyond the surface of a sphere concentric with the bubble. This may seem a rather unlikely result, but it has been verified, though only approximately, by experiment. The experiment (de Kock, 1961; Davidson, 1961) was to inject into a bed of lead shot, incipiently fluidised by water, a bubble of water containing dye, and then to observe how the dye and the bubble arrived at the surface. Plate IV shows two of the resulting photographs. Plate IVa shows how the disturbance of the surface takes place shortly after the first dye has reached the surface. The particles thrown up by the surfacing bubble appear as a dark area with a curved top, slightly to the right of the more transparent region due to the dye. In Plate IVb, the bed surface has become level again, and the cloud of dye, within a reasonably discrete volume, as predicted by the foregoing theory, is above the surface of the bed. The streamlines in fig. 28 are directly comparable with the experiment of Plate IV since u_0 and U_b were the same in both cases; the theoretical value of $A/a = 2\cdot22$ from (4.18(ii)) compares reasonably well with the observed value of $A/a = 1\cdot65$. Such agreement is all that can be expected in view of the fact that the theory is for irrotational wakeless flow of a spherical bubble; the experimental bubble must have been of the spherical-cap form with a following wake of fluidised particles. Mixing between the wake and the surrounding particulate phase presumably means that some of the fluidising fluid which left the rising bubble did not stay with it, and this probably accounts for some of the blurring of the edges of the dyed region in Plate IV b.

The experiment of Plate IV was chosen so that u_0 and U_b should be of the same order of magnitude. Plate V shows a similar experiment due to Rowe (1962 b), in which U_b/u_0 was between 2 and 3. A bubble of air containing NO_2 was blown into a two-dimensional air-fluidised bed of glass spheres of $\frac{1}{4}$ mm diameter. Plate V shows with striking clarity that the NO_2 remains within a circle concentric with the bubble, just as theory predicts; and the value of A/a from Plate V is certainly of the same order as the value given by (4.17(ii)).

In practical problems with gas-fluidised solids, the bubble rising velocity U_b is usually much greater than u_0. For example, with fine catalyst particles fluidised by gas, U_b might be ten times u_0, and then A/a from (4.18(ii)) is 1·1. The theory therefore predicts that the gas within a bubble should make contact only with a thin shell of particles surrounding the bubble. The contact will doubtless be improved by mixing within the wake below each bubble, but it seems certain that the fluid will penetrate only into a thin layer of particles above each bubble, because once the fluid enters the particulate phase, its motion is determined largely by the motion of the particles due to the rising of the bubble. It also seems likely that the order of magnitude of the exchange of fluidising fluid between a rising bubble and the particulate phase will be given by the theoretical equation (4.20); this result is used in Chapter 6.

4.5. Pressure distribution and inter-particle forces around a rising bubble

This section is devoted to a discussion of the pressure variation around a bubble rising in an incipiently fluidised bed of particles. The pressure P, defined in § 4.3 (a), p. 66, can be calculated from the particle velocities defined by (4.8), and the fluid pressure p_f can be calculated from (4.10), the discussion being confined to three-

(a)

(b)

Plate IV. Injection of a single bubble filled with coloured material. $U_b > u_o$. (a) A dyed water bubble breaking surface in a water-fluidised bed of lead shot. (b) The dye remains within a discrete volume.

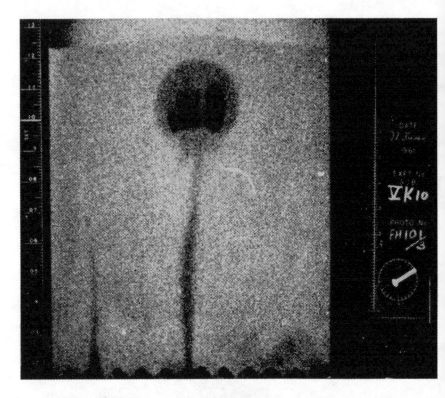

Plate V. Injection of a single bubble filled with coloured material. $U_b > u_0$. An air bubble containing NO_2 within an air-fluidised bed of glass beads; note that the NO_2 is confined to a circle concentric with the bubble (Rowe, 1962*b*, see also Rowe, 1962*a*).

dimensional motion. The difference between these two pressures then gives the inter-particle pressure p_p (4.7), and the magnitude of p_p determines whether or not the assumed theoretical model is reasonable; if p_p is large and positive it is unlikely that the particulate phase will behave as a fluid of zero viscosity; negative values of p_p are clearly impossible since they would imply tensile forces between the particles. Thus $p_p = 0$ is the only result fully consistent with the assumption that the particulate phase behaves as an incompressible fluid of zero viscosity.

The pressures are calculated along a vertical line through the centre of the bubble, for simplicity and because that is the region where the analysis is most likely to apply. From (4.8)

$$v_r = \partial\phi/\partial r = \mp U_b(1 - a^3/r^3)$$

when $\theta = 0$ or π, and with Bernoulli's theorem we get

$$\frac{P}{\rho_p} = B \mp gr - \frac{U_b^2}{2}\left(1 - \frac{a^3}{r^3}\right)^2, \tag{4.21}$$

where ρ_p is the average bulk density of the particulate phase and B is an arbitrary constant. Equation (4.21) applies only when the inertia of the particles is much greater than that of the fluidising fluid, so that the mass acceleration of the fluid can be ignored. The fluid velocity is of course different from the particle velocity, and inclusion of fluid inertia would greatly complicate (4.21); but in most bubbling fluidised systems, the particles are much denser than the fluid, so (4.21) is likely to be valid in most practical cases.

The pressure within the fluidising fluid is given by (4.10), J being determined from the boundary conditions at a great distance from the bubble, where the bed is incipiently fluidised by a uniform upward flow. Now in this condition it is an experimental fact that the pressure gradient within the fluid balances the weight of the particles, and therefore, for large values of r,

$$\left(\frac{\partial p_f}{\partial r}\right)_{\theta = 0 \text{ or } \pi} = \mp J = \mp \rho_p g, \tag{4.22}$$

and combining this with (4.10), with $\theta = 0$ or π, gives

$$\frac{p_f}{\rho_p} = \mp g\left(r - \frac{a^3}{r^2}\right). \tag{4.23}$$

It follows from (4.22) that $p_p = 0$ at a great distance from the bubble, because in that region the gradient of P must balance the weight of the particles, so that $\partial P/\partial r = \mp \rho_p g$ even if the bed is not fully fluidised. Hence from (4.7), $P = p_f$ for large r, and using this condition with (4.21) and (4.23) gives $B = \frac{1}{2}U_b^2$, and (4.21) becomes

$$\frac{P}{\rho_p} = \mp gr + \frac{U_b^2}{2}\left(\frac{2a^3}{r^3} - \frac{a^6}{r^6}\right). \qquad (4.24)$$

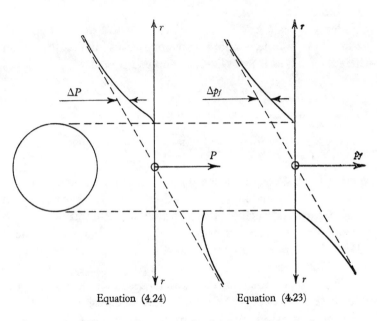

Equation (4.24)   Equation (4.23)

Fig. 29. Theoretical pressure distributions on the axis above a spherical bubble in a fluidised bed.

Fig. 29 shows the pressure distributions resulting from (4.23) and (4.24), the negative sign applying to the top half of the diagram and the positive sign to the bottom half, where there is an alternative. Fig. 29 shows that the pressures given by (4.24) must be incorrect below the bubble, because they do not satisfy the requirement that the pressure shall be constant throughout the bubble. Fig. 31 shows the pressure distribution likely to occur with an actual spherical-cap bubble, the distribution of P satisfying the requirement of constant pressure within the bubble. In the absence of any results for the spherical-cap bubble we shall use (4.23) and (4.24) to give some indication as to how the pressures vary *above* the bubble.

At the surface of the bubble, the inter-particle pressure p_p must be zero, and therefore from (4.7), (4.23) and (4.24) must give the same pressure at $r = a$. Hence

$$U_b = (2ga)^{\frac{1}{2}}, \qquad (4.25)$$

using the negative sign since we are considering the upper half of the bubble. Now the rate of rise of a spherical-cap bubble is $0 \cdot 711 (gD_e)^{\frac{1}{2}}$ (2.6) which is very similar to (4.25) provided D_e is put equal to $2a$; the factor $0 \cdot 711$ in (2.6) can then be compared with the factor unity in (4.25). This order of agreement is all that can be expected from the above theory, and it will therefore be assumed that (4.25) is satisfied; it is then easy to compare the pressure distribution of (4.23) with the pressures from (4.24) and the difference gives the magnitude of the inter-particle pressure p_p. Fig. 30 shows such a comparison, the quantities plotted being the actual pressures minus the gravity component due to the weight of the particulate phase. The two pressures P and p_f agree remarkably well at all values of r; of course they have been made to agree at large r, and at $r = a$ in view of (4.25), but the graphs show remarkable similarity at intermediate values, showing that p_p, the inter-particle pressure, is small at all points on the axis above the bubble.

Fig. 31 shows conjectural pressure distributions in the neighbourhood of a spherical-cap bubble, though points A and B are fixed by the analysis given in Appendix B. It is clearly possible for there to be a reasonably good agreement between P and p_f at all points in the bed, implying small values of p_p throughout the particulate phase. Recently Reuter (1963 a, b) has published details of experimental verifications of the theory of this chapter; he gives measurements of pressure profiles like those in fig. 31, and particle streamlines similar to those in fig. 28.

The questions set out in the introduction to this chapter can therefore be answered as follows.

(a) The roof of a rising bubble is prevented from falling in by fluidising fluid, which in flowing through the bubble exerts the necessary stabilising forces on the roof particles. In the same way, inertia forces generated by the motion of other particles round the bubble are balanced by the percolation of the fluidising fluid through the region round the bubble. Therefore each particle can be regarded as being in free motion without contact with its neighbours.

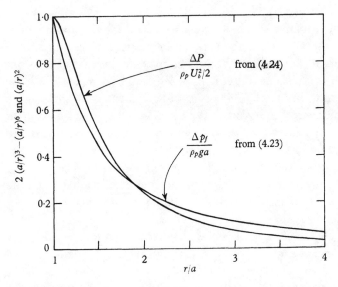

Fig. 30. Theoretical pressure distributions above a spherical bubble in a fluidised bed. These curves are derived from fig. 29 by subtracting the gravity components.

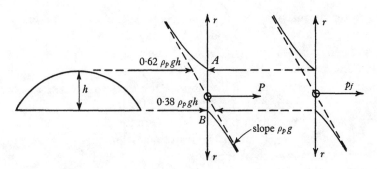

Fig. 31. Probable pressure distributions on the axis of a spherical-cap bubble in a fluidised bed.

(b) Since each particle exerts little force on its neighbours, the viscosity of the particulate phase can be expected to be small, and since there is no tendency for the particles to separate, the particulate phase should appear to be incompressible.

(c) Since the inter-particle forces are small, the only pressure to be considered is p_f, the pressure within the fluidising fluid.

The above analysis, though still incomplete and in need of further experimental proof, is largely independent of the size and nature of of the particles and of the properties of the fluidising fluid, and does therefore appear to provide sound reasons why bubbles in a fluidised bed behave almost exactly as if they were in an inviscid liquid.

With particular regard to (a) above, the theory of this chapter indicates that the roof of the bubble is in stable equilibrium. However, there is recent experimental evidence, from the observation of two-dimensional bubbles and X-ray pictures, that the equilibrium of the bubble roof is not always completely stable, and that 'curtains' or 'fingers' of particles from the roof can fall through the bubble as it rises. On occasions a bubble may be divided into two parts by the fall of particles. Very little of a quantitative nature is known about this phenomenon, but clearly it has considerable importance in the study of particle-fluid mass transfer in fluidised beds.

CHAPTER 5

THE STABILITY OF BUBBLES IN FLUIDISED BEDS

It was noted in Chapter 1 that when a bed of solid particles is fluidised by a gas the system normally exhibits bubbling and the fluidisation is said to be aggregative. On the other hand, when solid particles are fluidised by liquids the expansion of the bed is usually smooth and the fluidisation is described as particulate. However, there is now considerable evidence (Harrison, Davidson and de Kock, 1961; Simpson and Rodger, 1962) that no sharp demarcation exists between these two types of fluidised bed behaviour. This chapter considers how far it is possible to account for bubbling and non-bubbling systems in terms of the stability of the bubbles.

5.1. The transition between aggregative and particulate fluidisation

Simpson and Rodger (1962) have described the fluidisation of seven different solids (sand, three glasses, and three plastics) in turn by air, argon, arcton 6 (CCl_2F_2) and arcton 33 ($C_2Cl_2F_4$) over a range of pressures. Also, in a liquid-fluidised bed, they examined the fluidisation of glass beads, steel spheres and lead shot by water. These experiments were carried out in beds of about 3 in diameter. It was found possible to represent this wide range of gas- and liquid-fluidised systems by a single correlation involving voidage fraction, fluidising flow-rate, and the physical properties of the fluid and solids; and therefore, although the correlation is somewhat complex, it provides evidence that there is no fundamental difference between gas- and liquid-fluidised systems.

Several fluidised systems have been examined experimentally in order to illustrate the transition between aggregative and particulate behaviour, and the following are examples of the systems used.

Micro-balloon particles and carbon dioxide

Leung (1961) fluidised hollow phenolic micro-balloons with carbon dioxide at atmospheric pressure and observed bubbling.

75 % Glycerol   66 % Glycerol   39 % Glycerol   Water

$\epsilon = 0.60$

$\epsilon = 0.75$

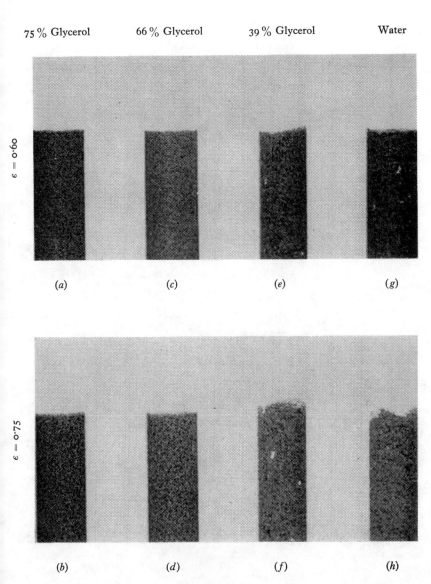

(a)   (c)   (e)   (g)

(b)   (d)   (f)   (h)

Plate VI. Fluidisation of lead shot with aqueous solutions of glycerol.

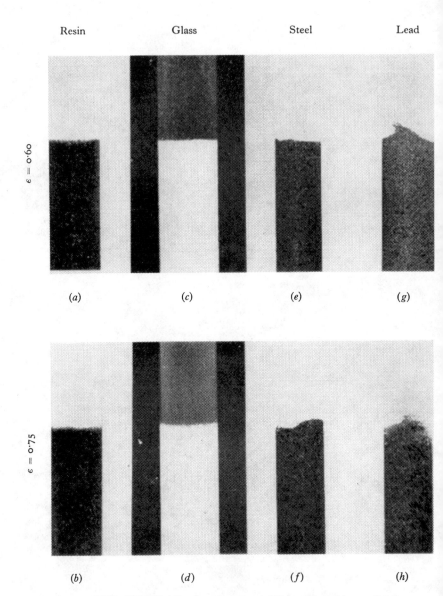

Resin   Glass   Steel   Lead

$\epsilon = 0.60$

(a)   (c)   (e)   (g)

$\epsilon = 0.75$

(b)   (d)   (f)   (h)

Plate VII. Fluidisation of various solid particles with paraffin.

However, when carbon dioxide at 600 lb/in² g. was used, the bed of particles fluidised smoothly. In between these two pressures there was a gradual transition. It may be noted that an increase of pressure from 0 to 600 lb/in² g. increases the density of carbon dioxide nearly 70 times.

Lead shot and aqueous solutions of glycerol

Plate VI shows the appearance of a bed of lead shot fluidised in a 1 in diameter bed by glycerol solutions of various concentrations (Harrison *et al.* 1961; de Kock, 1961); two voidage fractions are shown for each system. A transition from bubbling to smooth behaviour is observed as the concentration of glycerol increases. The 39, 66 and 75 wt. % solutions have, respectively, 3·5, 16·5 and 36 times the viscosity of water itself.

Particles of lead, steel, glass, and resin fluidised by paraffin

A transition in fluidised behaviour is observed (Harrison *et al.* 1961; de Kock, 1961) when paraffin is used to fluidise in turn particles of lead, steel, glass, and an ion-exchange resin. Plate VII illustrates the appearance of the bed at two voidages for each system. Whereas vigorous bubbling is observed in the lead shot–paraffin system, the resin–paraffin system behaves in a manner which is smooth and particulate. The particles of lead, steel, glass and resin have approximately the same size, but the respective particle densities are 11·32, 7·43, 2·90 and 1·50 g/ml.

These experiments on systems transitional in behaviour between aggregative and particulate illustrate the importance of the density ratio, $(\rho_s - \rho_f)/\rho_f$, and the fluid viscosity, μ: for instance, the fluidisation can be expected to become smoother as $\Delta\rho/\rho_f$ is decreased and μ is increased.

5.2. The stability of fluidised beds

An explanation of the origin of bubbles in fluidised beds is still lacking, although Rice and Wilhelm (1958) have shown that the lower surface of a bed supported merely by a gas stream is unstable. Their analysis predicts that a perturbation on the lower surface of a bed will tend to grow, and that this process is a function of the effective density and viscosity of the particulate phase. Investigations

6

carried out by Pigford (1959) and Jackson (1963) show, from rather similar premises, that very small disturbances about the state of uniform (truly particulate) fluidisation are initially amplified.

The full development of an infinitesimal departure from a steady state through to the fully grown void or bubble cannot be followed analytically, because the problem becomes intractable due to the size of non-linear terms in the equations of motion. Also, as Jackson points out, the importance of the non-linear terms is such that the fact that small disturbances begin to grow is no guarantee that the growth will be maintained to form true voids or bubbles. The problem can, however, be considered from another point of view, and that is the stability of the bubble once formed.

The stability of spherical-cap bubbles in fluidised beds

Plates VIII and IX show the fluidisation of lead shot, in the one case with air and the other case with water (de Kock, 1961). The appearance of injected bubbles of the fluidising fluid is quite different in the two cases. Injected bubbles of air in a bed of lead shot fluidised by air appear to be stable, the bubble being elongated if its equivalent diameter happens to exceed that of the containing vessel. On the other hand, bubbles of water in a bed of lead shot fluidised by water seem unstable and they appear to break up because solid material is gathered into the back of the bubble from the following wake. Each photograph illustrates a separate injection. It should be noted that, although these photographs certainly indicate some difference in behaviour between a gas-fluidised and a liquid-fluidised system, both are 'aggregative' in the normally accepted sense.

It is possible to find a partial interpretation of these experiments in terms of the analogy between the behaviour of fluidised beds and liquid–liquid or gas–liquid systems. This interpretation also throws some light on the nature of aggregative and particulate behaviour.

The circulation of fluid in a bubble. As a gas bubble rises in a liquid medium the shear force exerted by the liquid on the gas induces circulation of the gas within the bubble. The dotted lines in fig. 32 indicate the circulation pattern that might be expected in a spherical-cap bubble. A similar kind of circulation pattern of a liquid within a drop as the drop passes through a second liquid was observed by

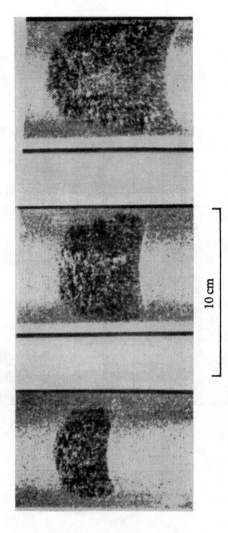

10 cm

Plate VIII. Fluidisation of 0·077 cm diameter lead shot with air, showing injected air bubbles.

(*Facing p.* 82)

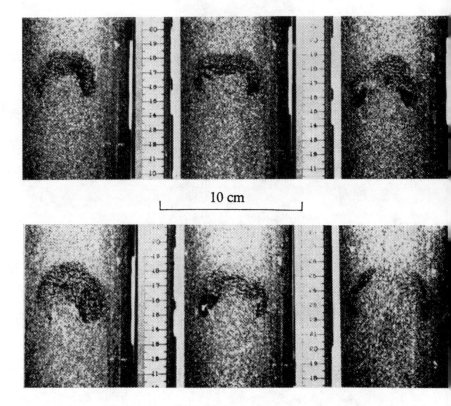

10 cm

Plate IX. Fluidisation of 0·077 cm diameter lead shot with water, showing injected water bubbles.

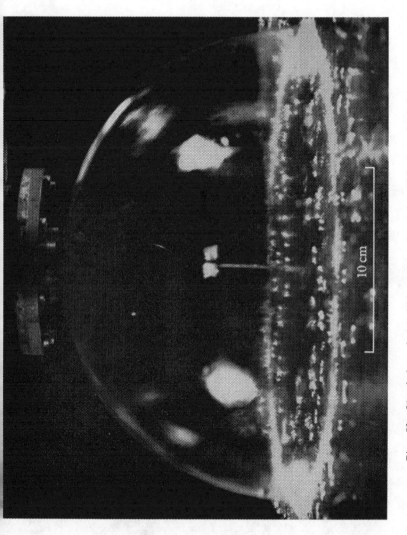

Plate X. Circulation of air within an experimental model of a spherical-cap bubble. The movement of air is demonstrated using an anemometer.

Garner and Haycock (1959). Moreover, it is possible to demonstrate circulation of gas in a spherical-cap bubble by means of the simple experiment photographed in Plate X. In this experiment a water nozzle was constructed to provide a thin sheet of water in the form of a spherical cap. The circulation of air within the 'bubble' so formed was then shown by means of a small anemometer. Actual measurements within an artificial bubble of this kind have been made by Rose (1961) and McWilliam (1961).

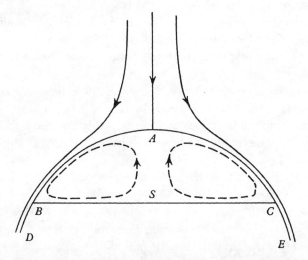

Fig. 32. Idealised flow patterns inside and outside a spherical-cap bubble held stationary by the downward flow of the continuous medium.

Fig. 32 illustrates a spherical-cap bubble in a fluidised bed, the bubble being imagined stationary with solid particles streaming downwards around it. The close relationship between two-phase systems and fluidised beds suggests that the downward motion of particles in the latter will bring about an effect analogous to the shearing action of one fluid on another in a two-phase system. The magnitude of the circulation so induced will be related to the maximum velocity of the stream of particles passing the bubble (i.e. at points B and C in fig. 32), and hence to the rising velocity of the bubble, U_b. However, the internal circulation pattern given in fig. 32 cannot be correct in detail, because it takes no account of the through-flow of fluid in the bubble in a fluidised bed discussed in

Chapter 4. Nevertheless, the experiments shown in Plates VIII and IX may be explained by supposing that there is an upward velocity, U_c, in the region S in fig. 32 which is of the same order of magnitude as the rising velocity of the bubble. For the purpose of the simple analysis of bubble stability which now follows it will be supposed that this upward flow arises as a combined consequence of internal circulation and fluid through-flow. This postulate is further considered and elaborated in § 5.5, but the crucial point of the simple hypothesis—that $U_c \approx U_b$—is not affected.

The stability condition for a bubble in a fluidised bed. Consider the possible behaviour of a solid particle which is transferred, by some means yet to be discussed (see § 5.5), from the wake to the inside of the bubble:

(i) If the upwards velocity, U_c, is equal to or greater than the free-falling velocity of the particle, U_t, there will be no tendency for the particle to fall back into the wake. In this case there will be a tendency for the bubble to 'fill-up' with particles from the wake (see Plate IX).

(ii) Alternatively, if the free-falling velocity of the particle exceeds U_c then it will fall back into the wake.

Now $U_c \approx U_b$, and thus a measure of U_c is given by (2.19). Moreover, if U_t is also known, the condition given by (i) and (ii), namely $U_t = U_b$, can be examined numerically.

It is easy to see in general terms that this condition predicts that gas bubbles in gas-fluidised systems will be stable to larger sizes than liquid bubbles in liquid-fluidised beds. U_b (and hence U_c) are approximately the same in gas- and liquid-fluidised beds provided bubbles of similar size are considered. However, the free-falling velocity of a particle in a gas often exceeds that of the same particle in a liquid by a factor of 1000, and therefore condition (i) above is usually reached at a smaller size of bubble for liquid systems than for gas systems.

5.3. The maximum size of stable bubble in a fluidised bed

In Chapter 2, (2.20) gave the rate of rise of a spherical-cap bubble of volume V in a liquid-fluidised bed. We shall here use the same form of equation but with the constant 0·71 replaced by 0·792 from the result of Davies and Taylor (1950), and with V replaced by $\frac{1}{6}\pi D_e^3$, so that

$$U_b = 0.711 g^{\frac{1}{2}} D_e^{\frac{1}{2}} \left[(1 - \epsilon_0) \bigg/ \left(\frac{\rho_s}{\rho_s - \rho_f} - \epsilon_0 \right) \right]^{\frac{1}{2}}, \qquad (5.1)$$

where ρ_f is the density of the bubble phase, and ρ_s is the density of the solid particles. The use of (5.1) in the analysis that follows implies the assumption that the excess fluid above that required for incipient fluidisation passes through the bed in the bubble phase.

The experimental value of the free-falling velocity, U_t, of a spherical particle of given diameter and density in a fluid of given density and viscosity may be readily evaluated using the tables compiled by Heywood (1962) from the results of a number of workers. The assumption that the particles are spherical is not of course valid in practice, but it is very doubtful whether a refinement of the theory to take account of particle shape would be justified at this stage.

In the first formulation of the theory of bubble stability (Harrison *et al.* 1961) the semi-empirical formula of Rubey (1933) was used to evaluate U_t. However, this formula is inaccurate for particle diameters greater than about 0.1 cm, and therefore in this chapter, U_b has been compared with the actual experimental values for U_t given by Heywood. The effect of this modification is to change some of the numerical results, but their general form is unchanged.

The bubble stability condition already discussed is such that when $U_b > U_t$ the bubble will retain solid particles that reach it from the wake. Therefore, the equality $U_b = U_t$ will give the criterion for the maximum stable size of bubble in a fluidised bed, V_m. This may also be characterised by the diameter of the equivalent sphere, such that $D_{em} = (6V_m/\pi)^{\frac{1}{3}}$. When U_b (from (5.1)) is made equal to U_t the values of D_{em} given in figs. 33 and 34 are obtained, respectively, for air- and water-fluidised beds. These graphs give D_{em}/d as a function of particle diameter, d, for various values of the solid–fluid density difference, $\rho_s - \rho_f \equiv \Delta\rho$.

It may be supposed that when D_{em}/d approximates to (or is less than) unity the size of the bubbles and particles will be comparable and the general appearance of the bed should be of smooth, or particulate, fluidisation. However, when D_{em}/d is greater than about 10 the system should exhibit bubbling, i.e. aggregative behaviour.

The régime of behaviour described by $D_{em}/d = 1 \to 10$ thus arbitrarily defines a transition from particulate to aggregative fluidisation.

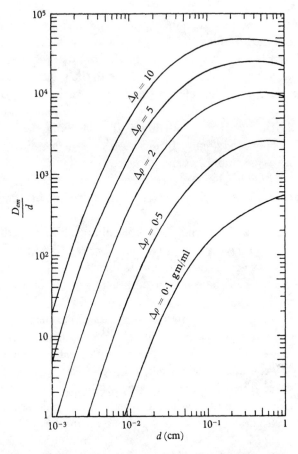

Fig. 33. Values of D_{em}/d for the fluidisation of solid particles with air. ($\rho_f = 1 \cdot 3 \times 10^{-3}$ g/ml; $\mu = 1 \cdot 8 \times 10^{-4}$ P; $\varepsilon_0 = 0 \cdot 40$.)

Figs. 33 and 34 predict the following characteristics for air- and water-fluidised beds:

(i) D_{em}/d for particles fluidised by air is large unless $\rho_s - \rho_f$ or d, or both, are small, and hence such systems should in general be aggregative.

(ii) D_{em}/d for particles fluidised by water is small unless $\rho_s - \rho_f$

and/or d are large, and hence particulate fluidisation may normally be expected.

(iii) In between the regions of typical bubbling and non-bubbling behaviour a transition from one type of fluidisation to the other should be observed.

The above theory of bubble stability will now be related to experiment.

5.4. The relation of the theory of bubble stability to experiment

The experimental evidence related qualitatively to the theory is extensive but, as will be explained, quantitative information is scarce.

(a) Qualitative aspects of the theory

The theory ascribes the differences in behaviour normally observed in gas-fluidised beds on the one hand and liquid-fluidised beds on the other to differences in bubble stability. In most gas-fluidised systems large bubbles are stable, whereas in most liquid-fluidised systems the largest stable bubble is comparable in size to the diameter of the solid particles, and thus the fluidised bed is smooth in appearance. An identical conclusion was reached on experimental grounds by Simpson and Rodger (1962) from their work on the fluidisation of light solids by gases under pressure and heavy solids by water.

When the size of the bubble is uninfluenced by the walls of the containing vessel, the analysis of bubble stability indicates that the size of the largest stable bubble is determined by ρ_s, ρ_f, d and μ. By varying these parameters it is possible to observe a complete range of behaviour from bubbling to smooth fluidisation in both gas- and liquid-fluidised systems. The manner in which the appearance of the bed depends on these variables will now be considered.

The density ratio $\Delta\rho/\rho_f$. As may be seen from figs. 33 and 34, the magnitude of $\Delta\rho/\rho_f$ is very important in defining the nature of the fluidisation for beds fluidised by air or water. The ratio $\Delta\rho/\rho_f$ certainly distinguishes between the available experimental results on gas-fluidised systems at atmospheric pressure ($\Delta\rho/\rho_f > 10^3$) and liquid-fluidised systems ($\Delta\rho/\rho_f < 10$): a table of data covering a wide variety of systems is given by Romero and Johanson (1962). Fluidisation can always be expected to become smoother as $\Delta\rho/\rho_f$ is reduced, as is demonstrated by the differing appearance in Plate VII of a bed of lead shot and a bed of resin particles when fluidised by paraffin.

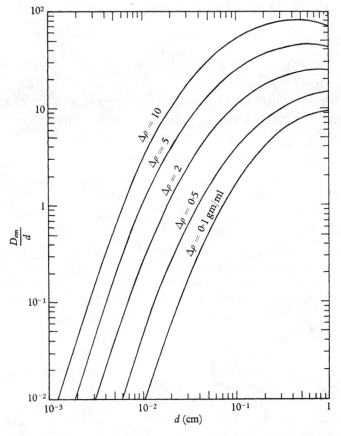

Fig. 34. Values of D_{em}/d for the fluidisation of solid particles with water. ($\rho_f = 1 \cdot 0$ g/ml; $\mu = 0 \cdot 01$ P; $\epsilon_0 = 0 \cdot 40$.)

The particle size. Figs. 33 and 34 show that particle size has a marked effect on the type of fluidisation. Smoother fluidisation can be expected when using finer particles, or when a proportion of fines is introduced into a bed of particles which would otherwise fluidise aggregatively. Work of Yasui (1956), reported by Romero and

Johanson (1962), has demonstrated particulate behaviour with the air-fluidisation of fine particles: hollow resin ($d = 0.008$ cm), and two sizes of catalyst ($d = 0.001$, 0.006 cm). However, in general, it is not easy to fluidise fine particles due to their tendency to agglomerate and bring about channelling of the fluid through the bed.

The viscosity of the fluidising fluid. The analysis of bubble stability indicates that the effect of the viscosity of the fluid on the type of fluidisation is greatest when the particle size is small. The effect of change of viscosity may be seen in Plate VI, which shows that the fluidisation becomes smoother the greater the concentration of glycerol in solution.

The fluidisation of lead shot and systems transitional in nature between aggregative and particulate mentioned in § 5.1 will now be considered in more detail.

The fluidisation of lead shot by air or water. The differing stability of bubbles in gas-fluidised and liquid-fluidised beds was noted in § 5.2 with reference to Plates VIII and IX which show lead shot fluidised in a 3 in diameter bed. The approximate values of D_{em}/d for these systems are: air-fluidised 40,000; water-fluidised 45. Hence, by using particles of 0.03 in diameter, air bubbles of some 100 ft may be expected to be stable when lead shot is fluidised by air; however, when water is used, the diameter of the largest stable water bubble should be no more than about 1 in. This is in line with the observation that when bubbles of a greater size are injected, as shown in Plate IX, they are unstable and tend to break up.

Micro-balloon particles and carbon dioxide. Table 5 shows that D_{em}/d decreases with increasing gas pressure, in qualitative agreement therefore with the transition from bubbling to non-bubbling fluidisation observed by Leung (1961).

Lead shot and aqueous solutions of glycerol. The change from bubbling to smooth fluidisation with increasing glycerol concentration is predicted by the decrease in the calculated values of D_{em}/d given in table 6.

Particles of lead, steel, glass and resin fluidised by paraffin. Plate VII shows that the lead shot–paraffin system exhibits bubbling, and this is consistent, with a calculated value of D_{em}/d of 38.0. This may also be contrasted in table 7 with $D_{em}/d = 1\cdot5$ for the particulate system of resin particles fluidised by paraffin.

Froude number criterion. The criterion that the Froude number ($\text{Fr} = U_0^2/gd$) is less than unity for most particulate systems and greater than unity for most aggregative systems was one of the earliest generalisations to come from a study of fluidised systems (Wilhelm and Kwauk, 1948). It is possible to relate this criterion to the theory of bubble stability.

Table 5. *Fluidisation of phenolic micro-balloons with carbon dioxide at various pressures*

($\rho_s = 0\cdot24$ g/ml; $d = 0\cdot0125$ cm)

CO_2 pressure (lb/in² g.)	ρ_f (g/ml)	μ (P × 10⁴)	$\dfrac{D_{em}}{d}$	Appearance of bed as ϵ is varied from 0·4 to 0·7
0	0·0016	1·48	26·6	Bubbling
100	0·0120	1·51	10·4 ⎫	Transition region
200	0·0275	1·54	8·1 ⎬	
400	0·0619	1·60	4·4 ⎭	
600	0·107	1·66	4·3	Smooth

Pinchbeck and Popper (1956) found that the ratio U_t/U_0 for fluidised beds varied over the range 10–75. Therefore, using the condition for bubble stability with this estimate of U_t, in (5.1)

$$(10 \rightarrow 75)\,U_0 = 0\cdot711 g^{\frac{1}{2}} D_{em}^{\frac{1}{2}} \left[(1 - \epsilon_0) \Big/ \left(\frac{\rho_s}{\rho_s - \rho_f} - \epsilon_0 \right) \right]^{\frac{1}{2}}$$

or

$$\frac{D_{em}}{d} = \frac{(200 \rightarrow 11000)}{\left[(1 - \epsilon_0) \Big/ \left(\dfrac{\rho_s}{\rho_s - \rho_f} - \epsilon_0 \right) \right]} \frac{U_0^2}{gd}.$$

This expression shows that D_{em}/d is proportional to the Froude number, the constant of proportionality depending very considerably on the properties of the system studied. However, this very approximate proportionality between D_{em}/d and Fr does suggest why Wilhelm and Kwauk had some success in using this dimensionless group as a guide to the nature of the fluidisation, and it seems likely that aggregative and particulate behaviour will be observed respectively with high and low values of the Froude number.

Table 6. *Fluidisation of lead shot with aqueous solutions of glycerol*

($\rho_s = 11\cdot32$ g/ml; $d = 0\cdot077$ cm)

% wt. glycerol	ρ_f (g/ml)	μ (P)	$\dfrac{D_{em}}{d}$	Appearance of bed as ϵ is varied, see Plate VI	
75	1·19	0·363	1·7	(a)	(b)
66	1·17	0·166	4·7	(c)	(d)
39	1·11	0·0358	20·2	(e)	(f)
0	1·00	0·0100	47·1	(g)	(h)

Table 7. *Fluidisation of various solid particles by paraffin*

($\rho_f = 0\cdot78$ g/ml; $\mu = 0\cdot02$ P)

Solid	ρ_s (g/ml)	d (cm)	$\dfrac{D_{em}}{d}$	Appearance of bed as ϵ is varied, see Plate VII	
Resin	1·50	0·060	1·5	(a)	(b)
Glass	2·90	0·0775	6·2	(c)	(d)
Steel	7·43	0·077	21·8	(e)	(f)
Lead	11·32	0·077	38·0	(g)	(h)

(b) Quantitative aspects of the theory

If a quantitative comparison between theory and experiment is to be made, it is very important to realise, first, that the theory predicts the *maximum* size of stable bubble to be expected in a given system. Thus to compare the predicted bubble size with *any* bubble size observed experimentally is no test of the theory, unless there are also reasons for believing that the experimental arrangement affords the bubbles full opportunity for growth (usually by coalescence) to a maximum size.

Secondly, it should be clear that the theory makes no claim to high accuracy for the precise numerical values of D_{em}/d, since the upwards velocity within the bubble, U_c, is equated as a rough approximation to the rising velocity of the bubble. The nature of the assumption that $U_c \approx U_b$ is considered further in § 5.5.

Quantitative information related to the theory falls into three categories:

(i) Evidence that when a bubble of a size exceeding the predicted maximum is injected into the bed it breaks up as it rises through the bed. This evidence has already been discussed in §§ 5.2 and 5.4 (*a*) with reference to the fluidisation of lead shot by air or water.

(ii) The observed sizes of bubbles in fluidised beds wide enough and deep enough to allow full and unimpeded bubble growth.

(iii) The observed sizes of bubbles in beds of very fine particles. Although in this case the fluidised bed may be of laboratory size, the experiments are often relevant to the theory because the predicted D_{em} may be small in comparison with the diameter of the bed.

An example of (ii) is the photograph by Hardebol (1961), shown in Plate XI, of air bubbles of about 16 in frontal diameter in a fluidised bed 10 ft high and 5 ft diameter. In a bed of such a size it is reasonable to suppose that the bubbles have had an opportunity for growth to the maximum size appropriate to the system. The photograph shows three bubbles bursting, or about to burst, at the bed surface, and it is particularly noteworthy that they all three have approximately the same frontal diameter. The theory predicts D_{em} to be 12 cm with $d = 0.02$ cm and $\Delta\rho \approx 1.5$, whereas the experimental value of D_e is 30 cm; this has been calculated from the frontal diameter, D_f, of the spherical-cap bubble by using (2.21) which gives, with $2\alpha_1 = 120°$, $D_f = 1.61 D_e$. The comparison between theory and experiment is therefore not good but, considering the approximations of the calculation, the correct order of magnitude is predicted.

When a bed of very fine catalyst particles ($d < 0.004$ cm) is fluidised by air at atmospheric pressure, bubbles of about 0.5–1 in may be observed in a bed of 1 in diameter and 6 in high. This bubble size is some 20 times the maximum diameter predicted by the theory (Lindsay, 1961; Crampton, 1961). Therefore, although qualitatively it is usually true to say that finer particles bring

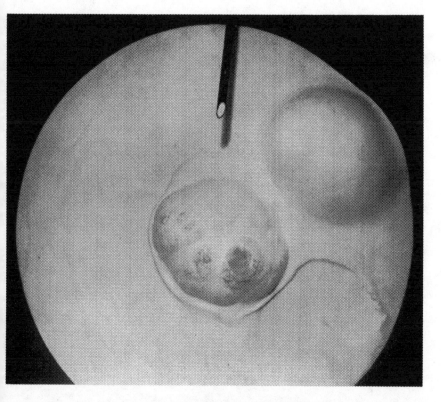

Plate XI. Plan view of catalyst bed fluidised by air (Hardebol, 1961).
Bed diameter 5 ft. Mean particle diameter 200 μ.

(Facing p. 92)

about smoother fluidisation, the simple theory of bubble stability fails to predict this trend quantitatively.

In any comparison between theory and experiment, in the way just described, some caution is necessary in estimating bubble diameter from the size of the disturbance caused at the surface of the bed. There is some evidence, mentioned in Chapter 2, that a considerable expansion of the bubble occurs at the moment the bubble surfaces, and therefore the bursting size may not be a good measure of the size of the bubble during most of its passage through the bed.

5.5. Further consideration of the theory of bubble stability

The simple theory of bubble stability that has been given in this chapter may be criticised on a number of grounds, for instance:

(i) It says nothing about the mechanism by which at the moment of instability (as shown in Plate IX) the particles leave the wake and fill up the bubble. It considers only the behaviour of a particle once in the bubble in relation to the possible flow of fluid within the bubble.

(ii) The free-falling velocity of an isolated particle is used in the calculations, whereas Plate IX certainly suggests that the particles are quite close together when they enter the bubble from the wake. The free-falling velocity of a cloud of particles is of course less than U_t, and this effect will therefore result in an over-estimate of D_{em}.

(iii) The calculation of U_b from (5.1), and hence each numerical value of D_{em}/d, implies that the bubble is spherical-capped whatever its size. This is indeed a questionable assumption when the bubble and particle sizes are similar. Not nearly enough is yet known about the shape and behaviour of very small bubbles in fluidised beds. In a gas–liquid system, surface tension ensures that small gas bubbles are nearly spherical. In a fluidised system this force is absent, and it is possible that the spherical-cap model is appropriate over a wider range of bubble sizes than would be possible in normal two-phase systems.

(iv) Rose (1961) and McWilliam (1961) measured the circulation of air within a water bell of the kind shown in Plate X. They found that the maximum upward air velocity on the axis of the bell was about one-quarter of the velocity U_b calculated by substituting the volume of the bell into (2.19). If this result applies to bubbles in a fluidised bed, such bubbles could be stable up to much larger sizes than is predicted by figs. 33 and 34.

(v) Although experimental evidence of a qualitative nature supports the theory, such quantitative evidence as exists is not in close agreement, especially for systems of very fine particles. It may be further argued that evidence of a qualitative kind is not particularly conclusive for, as Rowe (1962 a) has pointed out, the relevant parameters (ρ_s, ρ_f, μ and d) may be combined in various ways to distinguish between particulate and aggregative fluidisation.

Any extension or improvement of the simple theory must start by recognising that together with the circulation of fluid within the bubble there is also a through-flow. At the present time the interrelation between these two flows may only be conjectured, because experimental data on flow patterns within bubbles are lacking.

(a) The flow of fluid in a spherical void moving in an incipiently fluidised bed

In Chapter 4 the motion of the particles and the fluidising fluid outside a spherical void in an incipiently fluidised bed was examined in detail; and it may be noted that this analysis provides no information about the flow of fluid within the void itself. However, a stream function for flow within the void may be constructed if certain assumptions are made (Pyle and Rose, 1962), and this gives some indication of the interaction between the circulation of fluid within the void and the fluid through-flow. Fig. 35 (a)–(e) illustrates streamlines which satisfy the following conditions: (i) the radial component of the fluid velocity obeys continuity across the boundary of the void, and (ii) the tangential component of the fluid velocity does not change on crossing the boundary—a 'no slip' condition. For these calculations the particulate phase surrounding the void was assumed to have a voidage of 0·5.

This calculation indicates several different kinds of fluid movement, and two points of major interest may be noted:

(i) There are regions—shaded in fig. 35 (a)–(e)—of fluid circulation entirely confined to the void. The proportion of the volume of the void within this region of circulation increases as the ratio u_0/U_b decreases. There is always some fluid circulating within the void therefore, although the amount of circulation is very small when u_0/U_b is large.

(ii) An estimate of the magnitude of the local velocity at points B in fig. 35 (a)–(e) may be obtained. The total volume flow-rate across the diametral plane of the void normal to the main fluidising flow is shown in Chapter 4 to be $3U_0\pi a^2$ (4.20), where U_0 is the superficial fluidising velocity and a is the radius of the void. It

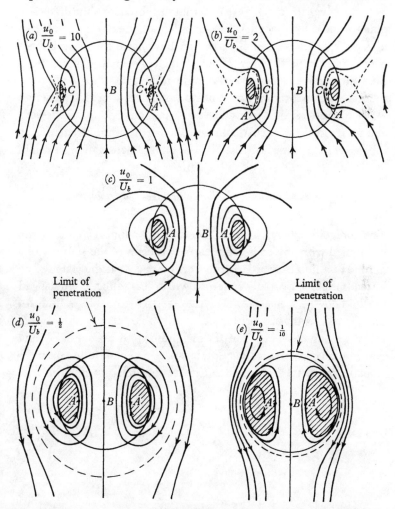

Fig. 35. Calculated flow patterns of fluid inside and outside a spherical void for various values of u_0/U_b (Pyle and Rose, 1962).

should be emphasised that this flow is across ABA in fig. 35 (a)–(e), and thus the local velocity at point B will always be greater than $3U_0$, especially when the regions of internal circulation in the void are extensive (e.g. fig. 35 (e)). In general it may be shown (Pyle and Rose, 1962) that the local velocity at points B is always greater than U_b, which is clearly relevant to the criterion of bubble stability put forward in §5.2.

It is noteworthy in fig. 35 (a)–(e) that the flow across the diametral plane of the void is shown to be dependent on the ratio u_0/U_b. When $u_0/U_b > 1$, the flow consists partly of a flow once through the bubble over CBC, and partly, over AC, of a fluid flow which circulates from the void to the particulate phase and back again. When $u_0/U_b < 1$, there is no true through-flow; the flow of magnitude $3U_0\pi a^2$ across ABA consists of a circulation of fluid on a path partly in the void and partly in the particulate phase. The extent of the circulation in the particulate phase is defined by the limit of penetration shown in fig. 35 (d), (e) and discussed in Chapter 4.

(b) *The stability of the lower surface of a bubble in a fluidised bed*

The foregoing analysis needs experimental support before it can be applied with any confidence to the problem of the stability of a bubble in a fluidised bed. Moreover, the fact that the shape of an actual bubble is not truly spherical will at least affect the details of the fluid flow patterns. However, there seem reasonable grounds for the hypothesis that in a spherical-cap bubble there is an upwards velocity along the centre line of a magnitude of the order of U_b, and that in the right circumstances this is large enough to lift particles from the wake into the bubble, so causing the break-up of the bubble.

CHAPTER 6

THE FLUIDISED BED AS A CATALYTIC REACTOR

6.1. Introduction

One of the most important applications of fluidisation is when a gas containing one or more reacting species fluidises a bed of particles which catalyse the desired chemical reactions of the gas. As the simplest case of this kind, consideration will be given to what happens when one of the components of the gas phase undergoes a first-order chemical reaction when it is in intimate contact with the catalyst particles of the fluidised bed; theoretical models will be constructed, based on the work described in the preceding chapters, and these models will be used to predict the degree of conversion with the first-order reacting system.

The foregoing work has shown that the fluidised bed consists of a particulate phase, agitated by rising bubbles of pure fluidising fluid, each bubble exchanging its fluid with the particulate phase as it rises. The initial size of the bubbles depends upon the distributor; if this is of the gauze type, the bubbles will be small initially, but will coalesce as they rise through the bed and utimately reach a limiting size whose order of magnitude is given by the theory of Chapter 5; if the distributor is of the bubble plate type, it may form bubbles of about the limiting size which will travel up the bed largely unchanged in size. The theoretical models described in this chapter assume that the bubbles are of uniform size throughout the bed, and that the catalytic reaction takes place entirely in the particulate phase; however, gas within the bubble phase cannot escape completely unreacted, because of the transfer between the phases, both by diffusion and by bulk flow of the type described in Chapter 4. These models lead to expressions for the overall degree of conversion due to reaction within the bed.

Similar models have been described by previous authors (Shen and Johnstone, 1955; Pansing, 1956; Mathis and Watson, 1956; Lewis, Gilliland and Glass, 1959; May, 1959; van Deemter, 1961; Massimilla and Johnstone, 1961) though without explicit reference to the mechanics of the bubbles; instead the results were variously interpreted in terms of a transfer coefficient or flow between the bubble and particulate phases. However, Zenz (1957*b*) gave a theory from which to calculate the amount of gas which failed to contact the particles; this theory is given by Zenz and Othmer (1960, p. 278), and is based on a consideration of the flow through a rising bubble. Orcutt *et al.* (1962) were the first to interpret catalytic reaction data in terms of bubble diameters, and this procedure will be followed in this chapter by calculating bubble diameters from the reaction data of the other workers. These bubble diameters D_e are shown in table 8, and although D_e does not obviously correlate with the bed properties, and is usually larger than the limiting value predicted by the theory of Chapter 5, there is a consistency in the values of D_e which suggests that it is a useful parameter with which to summarise reaction data. The results are particularly relevant when considering scale-up from laboratory apparatus to full-scale plant; it seems highly desirable that the value of D_e should be known in both the laboratory apparatus and in the full-scale plant.

Table 8 shows not only a consistency in the values of D_e, calculated from results with various reacting systems, but also in D_e obtained by more direct measurements. This provides further evidence for the correctness of the bubble model.

6.2.   Catalytic conversion with a first-order reaction

This section will describe two theoretical models for calculating the degree of conversion in a bed of catalyst particles fluidised by reacting gas. In both models it is assumed that all the gas in excess of what is required for incipient fluidisation passes through as bubbles which are of uniform size, each bubble exchanging gas with the particulate phase as it rises. It is assumed that there is no reaction within the bubbles, and that there is a first-order reaction, of known velocity constant, within the particulate phase. The reacting species is assumed to be transferred from the bubbles to the particulate phase by molecular diffusion and by a bulk flow calculated from the results given in Chapter 4. The difference between the two models is that in one case the particulate phase is taken to be perfectly mixed, and in the other case piston flow is assumed.

Table 8. *Bubble sizes in fluidised beds estimated in four ways*

     (a) Catalytic conversion of vapours in the bed.
     (b) Bed expansion data.
     (c) Bubble rising velocity.
     (d) Gas tracer experiment.

Author	Reaction	Particles Material	Diam. (μ)	Superficial velocity (ft/s)	Bed size Height (ft)	Diam. (in)	Method of estimation	Bubble diam. D_e (in)
Orcutt (1960)	Ozone decomposition	Silica	20–60	0.12–	1.03–	4, 6	(a)	1.3
		alumina		0.48	2.27		(a)	2.0
		treated			1.03		(b)	0.56–1.6
		with iron						
		oxide			2.27		(b)	3.8–6.1
Pansing (1956)	Combustion in catalyst regenerator	Alumina	44	0.26	1.6	10	(a)	1.8
			44	0.54	2.3			1.2
			79	0.32	1.6			1.2
			79	0.17	2.8			3.5
Mathis and Watson (1956)	Dealkylation of cumene	Alumina	74–147	0.4	0.0833– 1.0	2, 3, 4	(a)	0.33– 2.2
Massimilla and Johnstone (1961)	Ammonia oxidation	Alumina	105 average	0.26	0.64	4.5	(a)	2.9
May (1959)		Cracking catalyst	125	0.4–1.4	30	60	(d)	15
					30	60	(c)	6.7–12

Except for the link with the mechanics of bubbles, these models have already been described in the literature (Shen and Johnstone, 1955; Pansing, 1956; Mathis and Watson, 1956; Lewis *et al.* 1959; Massimilla and Johnstone, 1961). More complex models, allowing for diffusion in a vertical direction in both phases, have also been described (May, 1959; van Deemter, 1961). However, the work on bubble mechanics suggests that since each bubble is a discrete entity, diffusion from bubble to bubble should not be possible; diffusion within the particulate phase may also be an unimportant effect, as will be shown later, so that the models described here may be good enough for most practical purposes.

(a) Equations from bubble mechanics

In this chapter it will be necessary to use some of the relations derived in Chapter 2 for the mechanics of bubble swarms.

Since the fluid flow represented by $U - U_0$ is assumed to pass as bubbles through the particulate phase, then from continuity, as in (2.13),

$$NVU_A = U - U_0. \tag{6.1}$$

Since the bubbles cause the bed expansion from height H_0 to height H, (2.14) holds good, and

$$NVH = H - H_0. \tag{6.2}$$

The absolute bubble velocity U_A is assumed, as in (2.12), to be the sum of the natural rising velocity $0.711(gD_e)^{\frac{1}{2}}$ plus the upward velocity of the particulate phase between bubbles, $(U - U_0)$, so that

$$U_A = U - U_0 + 0.711(gD_e)^{\frac{1}{2}}. \tag{6.3}$$

Equation (2.17), which gives the bubble diameter in terms of bed expansion, can be derived from (6.1), (6.2) and (6.3) by eliminating NV and U_A.

(b) Perfect mixing in the particulate phase

Fig. 36 shows the symbols used in the analysis. The gas enters the bed with a concentration c_0 of reacting species; at height y the concentration within a bubble is c_b and this concentration is a function of y; within the particulate phase the gas contains a concentration c_p of reacting species, and c_p is independent of y because of the complete mixing. At the top of the bed, the gas leaving the particulate phase, with concentration c_p, is mixed with the gas leaving as bubbles, with concentration c_{bH}, and the two streams unite to produce a stream with concentration c_H. The latter stream has a flow-rate U per unit cross-sectional area of bed, and its component parts, the stream leaving the particulate phase and the stream leaving as bubbles, have flow-rates of U_0 and $U - U_0$, respectively.

A material balance on a single rising bubble gives the equation

$$(q + k_G S)(c_p - c_b) = V\frac{dc_b}{dt} = U_A V\frac{dc_b}{dy}, \tag{6.4}$$

where q is the volume flow-rate in and out of the bubble, S is its surface area, and k_G is the mass transfer coefficient between the main volume of the bubble, where the concentration is c_b, and the bubble walls, where the concentration is c_p. In (6.4), $U_A \, dt$ has been put equal to dy, because the changes within a single bubble rising with velocity U_A are being observed. Equation (6.4) may then be

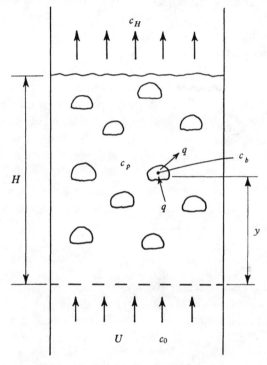

Fig. 36. Assumed bubble pattern in fluidised catalytic reactor, showing concentrations.

integrated with respect to y, using the boundary condition that $c_b = c_0$ at $y = 0$, and giving

$$c_b = c_p + (c_0 - c_p) e^{-Qy/U_A V}, \qquad (6.5)$$

where $$Q = q + k_G S. \qquad (6.6)$$

The quantity Q is an equivalent flow-rate to and from the bubble.

Since the particulate phase is assumed to be perfectly mixed, its material balance can be written down by considering the total height H with unit horizontal cross-section; there are five terms, as follows:

(i) The reactant entering the particulate phase from the bubbles, with flow-rate $NQ \int_0^H c_b \, dy$, where N is the number of bubbles per unit volume of bed. This term can be evaluated, using (6.5).

(ii) The reactant entering at the bottom, with a flow-rate $U_0 c_0$.

(iii) The reactant leaving the particulate phase and entering the bubbles, with a flow-rate $NQHc_p$.

(iv) The reactant leaving at the top of the bed, with a flow-rate $U_0 c_p$.

(v) The reactant consumed by the first-order reaction within the particulate phase. The rate of consumption within the region considered is $kHc_p(1 - NV)$, where k is the reaction velocity constant based on unit volume of the particulate phase; this unit volume is assumed to contain the same number of particles as unit volume of the bed at incipient fluidisation.

From the material balance on the particulate phase, the sum of terms (i) and (ii) must equal the sum of terms (iii) to (v), giving, after some rearrangement,

$$NVU_A(c_0 - c_p)(1 - e^{-QH/U_A V}) + U_0(c_0 - c_p) = kHc_p(1 - NV). \tag{6.7}$$

To obtain the overall conversion it is necessary to calculate c_H, the composition which results from mixing the streams leaving the bubble and particulate phases at the top of the bed, so that

$$Uc_H = (U - U_0)c_{bH} + U_0 c_p. \tag{6.8}$$

c_{bH} is obtained from (6.5) with $y = H$, and c_p is obtained from (6.7), and the expressions are substituted into (6.8); this is then rearranged, using (6.1) and (6.2) to eliminate NVU_A and NV, giving the following expression for the proportion of reactant which leaves unconverted

$$c' = \frac{c_H}{c_0} = \beta e^{-X} + \frac{(1 - \beta e^{-X})^2}{k' + 1 - \beta e^{-X}}, \tag{6.9}$$

where
$$\beta = 1 - U_0/U,$$
$$k' = kH_0/U$$
$$\left. \right\} \tag{6.10}$$

and
$$X = QH/U_A V.$$

Fig. 37 shows c' plotted as a function of k' for a fluidised bed, and this shows the effect of varying the dimensionless reaction velocity constant k' while keeping all the other variables, which affect the fluid mechanics, fixed. The most interesting feature of fig. 37 is that c' has a finite asymptote, $\beta\,e^{-X}$, at large values of k'. Thus, however fast the reaction, a proportion of the entering reactant manages to by-pass the bed, and this proportion depends only on the fluid mechanics of the bed, and not on the nature of the reaction. Such by-passing has been noted experimentally (Askins, Hinds and Kunreuther, 1951) in large fluidised beds of catalyst particles. Equation (6.9) shows that the by-passing is due to the bubbles, since the quantity X depends entirely on their properties.

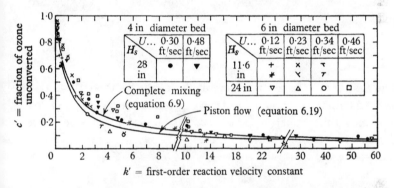

Fig. 37. Ozone conversion in a bed of catalyst particles fluidised by air. Data of Orcutt (1960). For the theoretical curves, $U/U_0 = 30$, $X = 2\cdot96$. (H_S = settled bed height.)

Other limiting cases give results that are more obvious than the by-passing effect, as follows:

(i) If $X \to \infty$, which implies a very high rate of exchange between the bubble and particulate phases, $c' \to 1/(1+k')$, the result for complete mixing within the whole bed.

(ii) If $k' \to 0$, $c' \to 1$, as it should.

(c) Piston flow in the particulate phase

A simple alternative to the above case is to assume piston flow in the particulate phase, so that the gas flowing through the particles is assumed to have a uniform concentration across a horizontal plane, but there is assumed to be no mixing in a vertical direction. Clearly this is an over-simplification, as was the assumption of complete mixing, and the truth must be that there is some mixing in the particulate phase, due to the agitation caused by the rising bubbles. Nevertheless, the two cases here presented cover the whole range of possible degrees of mixing of the particulate phase, and lead to simple analysis. More complex theory, allowing for eddy diffusion in the particulate phase, has been given (May, 1959; van Deemter, 1961).

With the assumption of piston flow, (6.4) still applies within each bubble, but (6.7) is replaced by a material balance on an infinitesimal height dy of the bed, either by considering a balance on the particulate phase, or on the whole bed within dy. The latter procedure gives a simpler result, namely

$$U_0\frac{dc_p}{dy}+(U-U_0)\frac{dc_b}{dy}+kc_p(1-NV) = 0. \qquad (6.11)$$

The first two terms are the incremental terms for the bubble and particulate phases, and the last term represents the consumption of reactant within the particulate phase. Equation (6.11) may be put in dimensionless form with the aid of (6.10); using (6.2) to eliminate NV we get

$$(1-\beta)\frac{dc_p}{dy}+\beta\frac{dc_b}{dy}+\frac{k'}{H}c_p = 0. \qquad (6.12)$$

The corresponding form of (6.4) is

$$\frac{dc_b}{dy}+\frac{X}{H}(c_b-c_p) = 0, \qquad (6.13)$$

and c_p can be eliminated from between (6.12) and (6.13), giving

$$H^2(1-\beta)\frac{d^2c_b}{dy^2}+H(X+k')\frac{dc_b}{dy}+k'Xc_b = 0. \qquad (6.14)$$

This linear second-order differential equation is solved by substituting

$$c_b = C_1 e^{-m_1 y}+C_2 e^{-m_2 y}, \qquad (6.15)$$

where m_1 and m_2 are the roots of the quadratic obtained from (6.14), so that

$$2H(1-\beta)m = (X+k') \pm [(X+k')^2 - 4k'X(1-\beta)]^{\frac{1}{2}}, \qquad (6.16)$$

where $m = m_1$ with the positive sign and $m = m_2$ with the negative sign. The arbitrary constants C_1 and C_2 are obtained from the boundary conditions at $y = 0$ which are that

(i) the fluid entering the bubble phase has the concentration c_0, and therefore at $y = 0$

$$c_b = c_0; \qquad (6.17)$$

(ii) and the fluid entering the particulate phase has the concentration c_0, so $c_p = c_0$, which, with (6.13) and (6.17), gives

$$\frac{dc_b}{dy} = 0 \qquad (6.18)$$

at $y = 0$. Substitution of the resulting values of C_1 and C_2 in (6.15) gives c_b as a function of y for substitution in (6.13) to give c_p as a function of y. From these expressions for c_b and c_p as functions of y, the concentrations at the top of the bed c_{bH} and c_{pH} are found by substituting $y = H$. Now (6.8) still applies in this case of piston flow, but with c_p replaced by c_{pH}, the concentration at the top of the particulate phase, and with the values of c_{bH} and c_{pH} obtained as described in the preceding sentence, we get the following expression for the fraction unconverted,

$$c' = \frac{c_H}{c_0} = \frac{1}{m_1 - m_2} \left[m_1 e^{-m_2 H} \left(1 - \frac{m_2 H U_0}{XU} \right) \right.$$
$$\left. - m_2 e^{-m_1 H} \left(1 - \frac{m_1 H U_0}{XU} \right) \right]. \qquad (6.19)$$

Fig. 37 shows c' plotted as a function of k' from (6.19) for given values of X and β. The corresponding graph obtained from (6.9) shows that, as would be expected, the 'piston-flow' model gives a better conversion than the 'complete-mixing' model. For large values of k', both models give the same fraction unconverted, $c' = \beta e^{-X}$, and this also is to be expected, because with a large value of k', all the reactant entering the particulate phase is completely converted, irrespective of whether there is piston flow or complete mixing in the particulate phase.

(d) Degree of conversion in terms of the equivalent bubble diameter D_e

The fraction unconverted, c', obtained either from (6.9) or (6.19), is a function of β, k' and X, which in turn are defined by (6.10). For a given fluidised bed in which there is a reaction of known velocity constant, β and k' are readily calculable; but X, depending as it does upon the equivalent flow Q through the bubble, defined by (6.6), is not so easy to find. It will be shown that X can be calculated from the equivalent bubble diameter D_e. Although D_e cannot be estimated from first principles, it will be apparent that its magnitude is a convenient summary of experimental results on conversion in any given fluidised bed.

To get X in terms of D_e, we eliminate Q from between (6.6) and (6.10), giving $X = (q + k_G S) H / U_A V$, and then get q from (4.20) with $a = \frac{1}{2} D_e$, and k_G from (C. 12), giving

$$X = \frac{H}{U_A V} \left(\frac{3 \pi U_0 D_e^2}{4} + \frac{0 \cdot 975 D_G^{\frac{1}{2}} g^{\frac{1}{4}} S}{D_e^{\frac{1}{4}}} \right). \tag{6.20}$$

This may be simplified by substituting $S = \pi D_e^2$, $V = \frac{1}{6} \pi D_e^3$, and by using (6.1), (6.2) and (6.3), eliminating NV and $U - U_0$, we get

$$\frac{H}{U_A} = \frac{H_0}{0 \cdot 711 (g D_e)^{\frac{1}{2}}}.$$

With these substitutions in (6.20), and some simplification, it becomes

$$X = \frac{6 \cdot 34 H_0}{D_e (g D_e)^{\frac{1}{2}}} \left(U_0 + \frac{1 \cdot 3 D_G^{\frac{1}{2}} g^{\frac{1}{4}}}{D_e^{\frac{1}{4}}} \right), \tag{6.21}$$

which can be used to calculate X for a given fluidised bed, if a value of D_e, the equivalent bubble diameter, is available. It must be borne in mind, however, that the theory leading to (6.21) does not have much experimental backing, particularly in regard to (i) the flow q between the bubble and the particulate phase, and (ii) the estimate of k_G, which is entirely theoretical, as there are no data on the gas film resistance between the main body of a rising bubble and its surface. Furthermore, no allowance has been made for the effect of the through-flow q on the magnitude of k_G.

6.3.   Comparison with experimental results for a first-order reaction

The following paragraphs describe the analysis of published experimental data for various catalytic reactions, using the above theoretical models as a basis for the analysis. The resulting estimates of bubble diameter D_e—shown in table 8—are in reasonable agreement with values estimated by other methods, and are consistent with visual observations of bubbling fluidised beds.

In every case described below, the catalyst bed was fluidised by a mixture of an inert with a reacting gas, and it was therefore necessary to estimate the value of the diffusion coefficient D_G for the mixture. These estimates were made using the method of Arnold or of Hirschfelder as set out by Reid and Sherwood (1958, pp. 267–8), and the results are shown in table 9. The importance of

Table 9. *Gas-phase diffusion coefficients (Reid and Sherwood, 1958)*

Method	Gases	P (atm)	T (°C)	$10^3 D_G$ (ft²/s × 10³)
Arnold	Ozone–air	1	50	0·22
	Cumene–benzene–propylene	1	511	0·20
	Ammonia–oxygen	1·10	250	0·74
Hirschfelder	Oxygen–air	1·69	551	0·74
	Oxygen–air	1·75	551	0·72
	Oxygen–air	2·91	551	0·43
	Oxygen–air	1·83	551	0·69
	Hydrogen–ethylene	1·01	113	0·98
	Helium–air	1	25	0·76
	CCl₄–air	1	20	0·078

these values is shown by the fact that the transfer of reacting substance between phases is mainly due to molecular diffusion, the bulk flow through the bubble being of secondary importance in all the cases shown in table 8.

(a) Ozone conversion (Orcutt, 1960)

The most satisfactory results for comparison with the above theory are due to Orcutt (1960), who measured the degree of conversion of ozone to oxygen in a mixture with air fluidising a bed of catalyst particles of mixed sizes, mainly in the range 20–60 μ. The particles were silica alumina, a proportion of which were impregnated with iron oxide. The value of the reaction velocity constant was varied, by varying the temperature of the bed, and was measured in a fixed bed of particles at the same temperature as the fluidised bed. The temperature range, from 80 to 190 °F, was such that k varied over a wide range of values. This meant that the degree of ozone conversion in the fluidised bed varied from a negligible amount to the asymptotic conversion at high values of k. This asymptotic conversion—which is shown by the presence of finite amounts of ozone in the stream leaving the bed, even at the highest values of k—is striking proof of the existence of by-passing. A similar effect was observed for a large catalyst regenerator by Askins *et al.* (1951); gas samples taken within the bed showed considerable variations in oxygen content, and the authors presumed that this was due to the sample tube drawing sometimes from the bubble phase—when the oxygen content was high—and sometimes from the particulate phase, when the oxygen content was low.

Orcutt presented his results in the form of graphs of c', the fraction of ozone unconverted, as a function of k', the dimensionless velocity constant. He found that the form of these curves is relatively insensitive to variations in

    (i) settled bed depth H_S, which ranged from 11·6 to 28 in,

    (ii) superficial air velocity U, which ranged from 0·12 to 0·48 ft/sec;

    (iii) bed diameter which was 4 or 6 in.

Because of this insensitivity, Orcutt calculated one value of X, the number of transfer units, for the whole series of experiments. This value of X was calculated by equating the theoretical asymptotic conversion, βe^{-X} from (6.9) or (6.19), to the observed asymptotic conversion.

Because of Orcutt's finding that one value of X will represent all his experimental results reasonably well, these results have been plotted on one graph, shown on fig. 37. Although there is a good deal of scatter, examination of the points shows that there is no systematic effect of H_S, U, or bed diameter. Following Orcutt's plan, one theoretical equation, (6.9), has therefore been fitted to all the points in fig. 37. In fitting (6.9) to the points, only one parameter, βe^{-X}, has to be fixed, and this was chosen as 0·05, which, with an average value of $U/U_0 = 30$ for all the points, gives $\beta = 0·967$ and $X = 2·96$. These are not very different from Orcutt's values of $\beta = 0·9$ and $X = 3·2$. Fig. 37 also shows the curve (6.19) for piston flow in the particulate phase, with $U/U_0 = 30$ and $X = 2·96$. Thus both curves in fig. 37 have the same asymptotic value of c' at high values of k', but the piston-flow model gives a better conversion at all values of k'.

An alternative procedure would have been to estimate X independently for the piston-flow model by fitting (6.19) to the experimental points in fig. 37. However, it seems clear that whatever value of X were chosen, (6.19) would not fit the results as well as (6.9), though there is not much to choose between the two equations; certainly on the basis of these results it is not possible to say whether piston flow or complete mixing in the particulate phase is more likely. Lewis *et al.* (1959) have made exactly the same observation.

Table 10. *Estimated bubble diameters for a fluidised bed of catalyst*
(*Orcutt, 1960; Orcutt* et al. *1962*)

Mixed particles mainly in the range 20–60 μ. Bed diameters 4 and 6 in

(1) From degree of ozone conversion (fig. 37 and equation (6.21) with $X = 2.96$)

H_0 (ft)	U_0 (ft/s)	D_e (in)
1.03	0.01	1.3
2.27	0.01	2.05

(2) From bed expansion; equation (2.17). Bed diameter 4 in

H_0 (ft)	U_0 (ft/s)	U (ft/s)	H (ft)	D_e (2.17) (in)
1.03	0.01	0.12	1.16	0.56
1.03	0.01	0.3	1.26	1.25
1.03	0.01	0.48	1.36	1.58
2.27	0.01	0.12	2.38	3.80
2.27	0.01	0.3	2.53	4.72
2.27	0.01	0.48	2.64	6.14

Estimation of a mean bubble diameter D_e for the bed. The value of $X = 2.96$ estimated from fig. 37 can be substituted in (6.21) which is then solved to give D_e, an estimated mean bubble diameter for the bed, and results are shown in table 10. In calculating D_e, the diffusion coefficient D_G was taken from table 9.

Table 10 also shows values of D_e estimated from the measured height H of Orcutt's catalyst bed at various values of the superficial air velocity U. These quantities were substituted into (2.17), which was derived by assuming that the bed contains bubbles of uniform size, all rising with the same velocity. In some cases the bubble diameters shown in part 2 of table 10 are larger than the bed diameter, indicating the possibility of slug flow which is caused by the bubbles filling the tube. Alternatively, it may be that the method of estimating bubble sizes from the measured bed height H is insufficiently accurate; with a bubbling bed, the fluctuations in bed height make it difficult to measure H, and (2.17) shows that the estimated bubble diameter D_e is highly sensitive to errors in H.

For comparison with table 10, table 11 shows bubble heights, for a similar range of particle sizes, measured by Yasui and Johanson (1958). The bottom line of table 11 refers to conditions which are quite close to those of table 10, in that the catalyst size and the values of U_0 and U, are about the same in both cases; it is noticeable that the bubble sizes are also very similar.

Table 11. *Bubble heights measured by Yasui and Johanson (1958)*
for 4 and 6 in diameter beds fluidised by air

Code	Material	Mean particle diam. (μ)	Particle density (g/ml)	U_0 (ft/s)	U (ft/s)	Bubble height (in)	Bed height (ft)
G 11	Glass beads	41	2·46	0·016	0·033	0·33	2·56
G 11		41	2·46	0·016	0·041	0·41	1·54
G 13		75	2·46	0·037	0·093	0·83	2·00
HR 4	Hollow resin	81	0·336	0·015	0·10	0·37	2·04
M 1	Magnetite	70	4·86	0·19	0·37	0·65	1·58
U 1	UOP catalyst	60	0·976	0·015	0·030	0·43	2·15
U 1		60	0·976	0·015	0·30	2·24	2·11

(b) Catalytic dealkylation of cumene (Mathis and Watson, 1956)

Mathis and Watson studied the dealkylation of cumene in a bed of 100–200 mesh (74–147 μ) silica alumina catalyst, the latter being fluidised by pure cumene at 950 °F in beds 2, 3 and 4 in diameter and up to 1 ft high. Their analysis of the data in terms of a two-phase model, assuming piston flow in each phase, led to a transfer coefficient K_{AB} defined as

$$\frac{\text{gram moles transferred between phases}}{\text{(h) (ft}^3 \text{ of bed) } (\Delta x \text{ between phases)}} \cdot$$

Here x is the moles of cumene converted, divided by the moles fed at the base of the reactor. Greensfelder, Voge and Good (1945) showed that the products of reaction are almost exclusively benzene and propylene, so that each mole of cumene yields, when converted, two moles of product gases. In the experiment of Mathis and Watson, the conversion of cumene averaged about 50 %, and under these conditions, using the notation of this chapter, $K_{AB} = QNC_0$ approximately, C_0 being the concentration, in moles per unit volume, of cumene entering the bed. From (4.20) with $a = \frac{1}{2}D_e$, (6.1) with $V = \frac{1}{6}\pi D_e^3$, (6.3) and (6.6) with $S = \pi D_e^2$, and (C.12) we can, by eliminating q, N, U_A and k_G, get

$$\frac{K_{AB}}{3600} = \frac{6C_0[0\cdot975D_G^{\frac{1}{2}}g^{\frac{1}{4}}/D_e^{\frac{1}{4}}+\frac{3}{4}U_0]}{D_e[1+0\cdot711(gD_e)^{\frac{1}{2}}/(U-U_0)]}, \tag{6.22}$$

Table 12. *Bubble sizes estimated from the catalytic dealkylation of cumene (Mathis and Watson, 1956)*

Bed size			
Height (ft)	Diam. (in)	K_{AB}	D_e (in)
0·083	2	1,500	1·2
0·167	2	3,000	0·80
0·5	2	2,000	1·0
1·0	2	900	1·7
0·125	3	12,000	0·33
0·25	3	2,000	1·0
0·75	3	900	1·7
0·33	4	1,000	1·6
1·0	4	600	2·2

and using this formula, D_e can be calculated from the results of Mathis and Watson. They found that K_{AB} varies considerably with the velocity U, having a peak value at $U = 0\cdot4$ ft/sec, and these peak values were used in calculating D_e from (6.22). From the paper of Mathis and Watson it is not clear why K_{AB} should reach a peak value at $U = 0\cdot4$ ft/sec, although perhaps there is incomplete fluidisation at lower velocities and slugging at higher velocities.

Table 12 shows equivalent bubble diameters calculated from (6.22), using $U_0 = 0\cdot02$ ft/sec, $C_0 = 0\cdot441$ g mole ft^{-3}, and from table 9, $D_G = 0\cdot20 \times 10^{-3}$ ft^2 sec^{-1}. D_G was estimated by Arnold's method, as set out by Reid and Sherwood (1958, p. 267), for cumene diffusing through a mixture of benzene and propylene, the principal products of reaction, the three species being assumed to be present in equimolar proportions.

The results in table 12 show a good deal of scatter, but the values of bubble diameter are consistent with the visual appearance of this type of bed in the laboratory and, in particular, it is striking that the bubble diameters tend to increase with bed height. Once again it is clear that with beds taller than about 1 ft, the bubble diameter is likely to exceed 2 in so that slug flow will occur in small diameter beds leading to behaviour quite different from that of a large reactor. Some of the results shown in table 12 are summarized in table 8 and will be further discussed.

(c) Combustion in a catalyst regenerator (Pansing, 1956)

Pansing gave data from pilot plant tests on the regenerator of a catalytic cracking unit. The air which fluidised the spent catalyst supplied oxygen to burn off the carbon on the catalyst, and Pansing interpreted his results in terms of a first-order velocity constant for the reaction of the oxygen within the particulate phase, and a mass transfer coefficient k_d between the bubble and particulate phases. He gave the empirical formula

$$k_d = \frac{\rho_f^2 U^2}{146d^{1.5}} \text{lb moles oxygen (h atm lb catalyst)}^{-1}, \quad (6.23)$$

where $\rho_f U$ lb ft^{-2} h^{-1} is the superficial mass velocity through the bed, and d microns is the particle diameter. By using the theory set out in this chapter, it is possible to calculate an equivalent bubble diameter D_e from (6.23) for any particular operating condition.

In unit volume of fluidised bed, the rate of transfer of oxygen between the bubbles and the particles due to a concentration difference $c_b - c_p$ is $NQ(c_b - c_p)$, where $c_b - c_p = \Delta p/RT$. The weight of catalyst per unit bed volume is $\rho_s(1 - \epsilon_0)(1 - NV)$, and therefore the rate of oxygen transfer per unit weight of catalyst is

$$k_d \Delta p = \frac{NQ\Delta p}{RT\rho_s(1 - \epsilon_0)(1 - NV)}.$$

This equation may be simplified by using (6.1) and (6.3) to eliminate N and U_A, giving

$$k_d = \frac{Q(U - U_0)}{VRT\rho_s(1 - \epsilon_0)\, 0.711(gD_e)^{\frac{1}{2}}},$$

and this becomes

$$\frac{k_d}{3600} = \frac{U - U_0}{RT\rho_s(1 - \epsilon_0)\,0{\cdot}711(gD_e)^{\frac{1}{2}}}\left(\frac{9U_0}{2D_e} + \frac{6 \times 0{\cdot}975 D_G^{\frac{1}{2}} g^{\frac{1}{4}}}{D_e^{\frac{5}{4}}}\right). \quad (6.24)$$

In the derivation of (6.24), $V = \frac{1}{6}\pi D_e^3$, and to derive Q, (6.6) was used with $S = \pi D_e^2$, (4.20) with $a = \frac{1}{2}D_e$, and k_G was obtained from (C. 12). Values of D_e, shown in table 8, were calculated from (6.24) for a range of the variables covered by Pansing; the particle sizes and bed heights are shown in table 8, and the estimated diffusion coefficients in table 9. This table also gives the pressures and temperatures for Pansing's experiments in the same order as the data in table 8. R was taken as $1{\cdot}31$ atm ft³/lb mole °K. It was necessary to assume that for the alumina catalyst, $\rho_s(1 - \epsilon_0) = 55$ lb/ft³, this being the bulk density at incipient fluidisation. It was also necessary to assume values for U_0 and these were estimated from the data of May (1959), who used similar particles; the values used were $U_0 = 0{\cdot}005$ ft/sec with $d = 44\,\mu$, and $U_0 = 0{\cdot}03$ ft/sec with $d = 79\,\mu$.

The bubble diameters estimated from (6.24) and shown in table 8 are what might be expected from visual observation of a 10 in diameter bed of small particles. They are also reasonably consistent with the bubble diameters estimated for smaller beds, suggesting that the bubbles do reach a limiting size in quantitative agreement with the theory of Chapter 5.

(d) Catalytic hydrogenation of ethylene (Lewis et al. 1959)

Lewis et al. (1959) measured the rate of hydrogenation of ethylene in a 2 in diameter fluidised bed of alumina catalyst. The ethylene was present in excess, so that the reaction was essentially first order with respect to hydrogen concentration when the gas was in contact with the catalyst. The authors analysed their results in terms of theoretical models similar to those described in this chapter; their 'vertically unmixed emulsion' model is similar to that of §6.2 (c), p. 103; and their 'completely mixed emulsion' model is similar to that of §6.2 (b), p. 100. Lewis et al. (1959) concluded, as did Orcutt (1960) for another system, that it was impossible to distinguish between these two models on the basis of their data on ethylene hydrogenation. They deduced a quantity F which represents the interchange of

8

gas between the bubble and particulate phases, in the units of (ft³ of gas) (sec)⁻¹ (ft³ of quiescent emulsion phase)⁻¹. With the notation of this chapter, $F = QNH/H_0$, and using (6.1) to eliminate N, then with (6.10),

$$F = X(U - U_0)/H_0,$$

and this, with (6.21), gives a means of calculating D_e from the measured values of Lewis *et al.* (1959) of F. In a typical case, $U = 0.5$ ft/sec, $U_0 = 0.024$ ft/sec, $F = 0.55$ for 122 μ particles, and taking $D_G = 0.98 \times 10^{-3}$ ft²/sec from table 9 gives $D_e = 3.4$ in. This is of course a bubble diameter larger than the column diameter of 2 in, which suggests that the fluidised bed was operating in the slug flow régime; for a 2 in diameter bed about 1 ft high, slug flow is to be expected at a superficial velocity of 0.5 ft/sec. Moreover, bubble diameters of about 3 in are consistent with the results shown in table 8, and with the measured bubble height shown in the last line of table 11.

(e) Oxidation of ammonia (Massimilla and Johnstone, 1961)

Massimilla and Johnstone fluidised a 4.5 in diameter bed of alumina catalyst (100–325 mesh; average particle size 105 μ) with gas consisting of 90 % oxygen and 10 % ammonia at 250 °C and 1.1 atm. Under these conditions the oxidation reaction is pseudo-first order, and the measured conversion of ammonia in the fluidised was compared with the measured conversion in a fixed bed under the same conditions of pressure and temperature. The results were interpreted in terms of the theory of §6.2 (c), p. 103, which assumes piston flow in the particulate phase. Comparing Massimilla and Johnstone's equations with equations (6.4) and (6.11) it can be shown that their transfer coefficient between phases, K_d min⁻¹, is given by

$$\frac{K_d}{60} = \frac{Q}{V} = \frac{9U_0}{2D_e} + \frac{5.85 D_G^{\frac{1}{2}} g^{\frac{1}{4}}}{D_e^{\frac{5}{4}}}. \tag{6.25}$$

In the derivation of (6.25), $V = \frac{1}{6}\pi D_e^3$, and to get Q, (6.6) was used with $S = \pi D_e^2$, (4.20) with $a = \frac{1}{2}D_e$, and k_G was obtained from (C. 12) as for (6.24). Massimilla and Johnstone give K_d as a function of gas flow-rate for bed depths of 0.64, 1.28 and 1.91 ft; it is striking that K_d diminishes sharply with increasing bed height, indicating a progressive increase in bubble size with bed height. By substituting the measured values of K_d into (6.25) it is easy to calculate the equivalent bubble diameter D_e. For a bed depth of 0·64 ft, fig. 6 of Massimilla and Johnstone gives $K_d = 185\,\text{min}^{-1}$ at a gas flow of 1 standard ft³ min⁻¹, corresponding to $U = 0.26\,\text{ft/sec}$. Equation (6.25) gives the value $D_e = 2.9$ in as shown in table 8; in calculating this value, D_G was taken from table 9. Massimilla and Johnstone do not give the incipient fluidising velocity U_0 and this was therefore estimated from their Fig. 6 which shows K_d tending to zero at a gas flow of 0·17 standard ft³ min⁻¹. This was presumed to be the point of incipient fluidisation, giving $U_0 = 0.045\,\text{ft sec}^{-1}$, a reasonable result for the particles used, and fortunately the U_0 term in (6.25) is relatively small.

With a bed depth of 1·28 ft the bubble diameter, similarly estimated, is $D_e = 7$ in, indicating the occurrence of slug flow, and a similar result would be obtained for the deeper bed of depth 1·91 ft. The results of Massimilla and Johnstone thus provide further evidence that in fluidised beds less than 1 ft deep the bubble size is of the order of an inch or two, but that with greater depths, the bubble size may be much larger, leading to slug flow in all but the largest laboratory apparatus.

Massimilla and Johnstone give an interesting commentary on the similar earlier work of Shen and Johnstone (1955) who studied the catalytic decomposition of nitrous oxide in a fluidised catalyst bed. The latter results give very large bubble diameters when interpreted by the theory of this chapter, and are also somewhat inconsistent with the results of Massimilla and Johnstone and with those of Mathis and Watson (1956); however, Massimilla and Johnstone list four reasons why the results of Shen and Johnstone should be treated with reserve and they have therefore not been further analysed.

(f) Helium injection experiments (May, 1959)

In his paper on large fluidised beds, May gives a theory, similar to the above, for a first-order reacting system, although no experimental results are quoted. However, results are given from transient experiments in which the helium supply to the inlet air stream of a large fluidised bed was suddenly cut off, and the helium concentration at the air outlet was subsequently observed as a function of time. The theory of this method required the solution of transient forms of (6.4) and (6.7), and (6.12) and (6.13) that is, with $\partial c/\partial t$ terms added, and these solutions were obtained on a digital computer. May gives a good deal of interesting information for a fluidised reactor 30 ft high and 5 ft diameter, but unfortunately no single consistent set of data are given, and to make deductions it is necessary to resort to guesswork to supply the information that is lacking. Nevertheless, in view of the lack of data on large fluidised beds, it is worthwhile to make some estimate of the bubble sizes in May's equipment, and this can be done in two ways, namely

(i) from the results of the helium injection experiments,

(ii) from the quoted bubble velocities deduced (it is supposed) from bed expansion measurements.

(i) May gives a series of computer results (his Figs. 13–15) which relate the conclusions of a helium tracer study to the so-called 'cross-flow ratio', which is the total rate of exchange of gas between the bubble and particulate phases, W_s, divided by the flow-rate of gas as bubbles, $U - U_0$. In the notation of this chapter, $W_s = NQH$, and using (6.1) to eliminate N, then

$$\frac{W_s}{U - U_0} = \frac{QH}{U_A V} = X, \qquad (6.26)$$

the quantity used above (6.10) in analysing Orcutt's (1960) results. Now from May's Figs. 13–15 and from his Table 3 it can be inferred that the cross-flow ratio is a small integer, and since U ranges from 0·4 to 1·4 in May's Figs. 13–15, a value of $X = 3$ will be assumed. Inserting this value in (6.21) and solving for D_e, with $H_0 = 30$ ft, $U_0 = 0$·05 ft/sec and $D_G = 0$·76 × 10^{-3} ft²/sec for helium in air, from table 9, gives $D_e = 1$·29 ft, a reasonable value for a 5 ft diameter bed, 30 ft high.

(ii) May's Figs. 13–15 give 'bubble gas velocity' which can be presumed to be the rising velocity U_A of the gas bubbles, and further, presuming that these data refer to large reactors, we calculate the bubble diameters from (6.3), giving

$$D_e = 0\text{·}56,\ 0\text{·}87\ \text{and}\ 1\text{·}02\ \text{ft},$$

which agree reasonably well with the value of 1·25 ft calculated from the cross-flow ratio.

It is possible to compare these bubble sizes with the limiting sizes predicted from the theory of Chapter 5. If from May's Table 1 the catalyst size is taken to be 125 μ, and the solid density of the catalyst is assumed to be 2 g/ml, then the theory of Chapter 5 gives a value of D_{em}/d of 650, leading to an estimated bubble diameter of 0·27 ft. This is less than the diameters estimated from May's data, thus confirming that the theory of Chapter 5 tends to predict bubble diameters which are too small.

(g) Injection of bubbles containing carbon tetrachloride (Szekely, 1962)

Szekely has described experiments in which single bubbles of air containing carbon tetrachloride were injected into a bed of silica alumina catalyst incipiently fluidised by a steady flow of pure air. The bubbles were injected through a tube immersed in the catalyst, and there was a time interval of about 3·4 sec between bubbles, each bubble being suddenly injected by briefly opening a valve connected to a high-pressure air supply. With this arrangement it could be assumed that the bubbles did not interfere with one another. By measuring C_0, the steady concentration in the outlet air stream, it was possible to deduce the amount of carbon tetrachloride which an individual bubble carried through the bed, the concentration in equilibrium with the particles being negligible. Szekely calculated C_i the concentration which would have resulted from mixing the injected mixture with the fluidising air stream; thus C_0/C_i was the fraction by-passing the bed, and this was given for a variety of bubble sizes and bed heights.

These results are easy to compare with the foregoing theory; the relevant equation is (6.5) with $c_p = 0$, the carbon tetrachloride being completely adsorbed in the particulate phase. The concentration within the bubble is then $c_b = c_0 e^{-Qy/U_b V}$, U_b replacing U_A, since each bubble is rising through the stagnant particulate phase; c_b and c_0 are not the same as C_0 and C_i, the latter being the concentrations when the bubble stream is diluted with the fluidising air, but since the dilution is the same at inlet and outlet,

$$\frac{C_0}{C_i} = \left(\frac{c_b}{c_0}\right)_{y=H_0} = e^{-QH_0/U_b V} = e^{-X}, \qquad (6.27)$$

where X is given by (6.21).

Fig. 38 shows the comparison between Szekely's results and (6.27); clearly the experimentally measured by-passing is much less than the theoretical, about one-third as much in the majority of cases. The most likely explanation for the differences is that they are due to end effects, and Szekely concluded that most of the adsorption takes place when the bubbles are formed. It is also possible that there was a considerable wall effect in the experiments because the average value of D_e was about 4·5 cm, and the bed diameter was 9·52 cm.

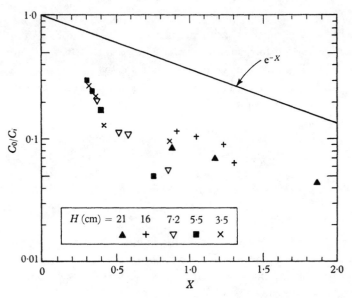

Fig. 38. Adsorption of carbon tetrachloride from a single bubble injected into a fluidised bed (Szekely, 1962).

It should be noted that in calculating the quantity X from Szekely's results it was necessary to estimate $U_0 = 0.0118$ ft/sec from (1.17), for Szekely's 60 μ particles, assuming their solid density was twice that of water. The value of D_G was taken from table 9.

6.4. Discussion of the theoretical models

The bubble sizes D_e shown in table 8 are reasonably consistent with one another and with visual observations of fluidised beds; likewise the agreement between theory and experiment shown in fig. 37 and the consistency of the bubble sizes shown in tables 10 and 11 is remarkably good, in view of the gross over-simplifications of the theory. However, in all these cases, the bubble sizes are considerably larger than the limiting values given by the theory of Chapter 5. There does not, therefore, seem to be any prospect at present of using the theory of Chapter 5 to predict an average bubble size D_e from which to predict the overall conversion in a catalytic reaction. But the calculation of D_e from laboratory and pilot plant results, as in table 8, does seem a useful way of summarising experimental results from catalytic converters, and these values of D_e should be valuable in predicting the performance of scaled-up plant. This will be further discussed.

In applying the theory of this chapter, particularly to large fluidised beds for which so few data are available, the following deficiences should be borne in mind.

(a) In any particular bed, the theory assumes that the bubble size is constant, though Yasui and Johanson (1958) found that in laboratory fluidised beds the bubble size increases with height above the support. This finding is in line with the results in tables 8, 10 and 12, where the bubble size is consistently larger for taller beds; moreover, the fact of fitting the theoretical curves in fig. 37 by using only one value of X leads, from (6.21), to the result that D_e must be approximately proportional to $H_0^{\frac{4}{7}}$, neglecting the U_0 term. This fact—that bubble sizes increase with bed height from a small value just above the support—must mean that near the support the gas-particle contacting is very good, leading to high conversion in this region. This effect is more important at very high values of k' for which a considerable degree of conversion may take place just above the bed support. Therefore more emphasis might have been given to the lower and intermediate values of k' in fitting the theoretical curves to the results shown in fig. 37. This would have led to smaller values of X, and hence larger bubble diameters, giving more satisfactory agreement between methods (a) and (b) in the first part of table 8. This point will be referred to later in the section dealing with mass transfer.

(b) It was emphasised in the theoretical derivation that the values of k_G for each bubble, and of the flow through it, are highly tentative; thus the theoretical derivation of (C. 12) takes no account of transfer to the lower part of the bubble, and is not backed by experimental evidence. However, the theory is to some extent supported by the reasonable consistency of the values of D_e shown in table 8. It is noteworthy that in every case shown in table 8, the transport of reacting substance between the bubble and particulate phases was largely due to molecular diffusion to the wall of each bubble, the effect of flow through the bubbles being small by comparison. With very large bubbles, the molecular diffusion term would be slightly less important, due to the $1/D_e^{\frac{1}{4}}$ term in (6.21), and with large particles the U_0 term would be more important.

(c) The theory assumed that the rising velocity U_b of a cloud of bubbles in a fluidised bed is the same as for a single bubble. In practice the bubbles must interact, and this will both alter the rising velocity and affect the interchange of gas between the bubble and particulate phases.

(d) In many cases the bubble diameters shown in table 8 are not much less than the diameter of the bed, though the theory was expressly for small bubbles in an infinite medium.

Application to large fluidised beds; the scale-up problem

It is interesting to note that the deficiencies (a) and (d) above are more serious for a small bed than for a large bed.

Thus effect (a) is due to the coalescence of bubbles just above the support plate, and the region of coalescence is likely to occupy a smaller proportion of a tall bed. This is confirmed to some extent by the results in table 8 which show that bubble sizes of about 2–3 in are reached within a bed height of about 1 ft, whereas May's data show bubble sizes of 6–15 in with a bed height of 30 ft.

Similarly deficiency (d), due to the finite ratio of bubble to bed diameter, is likely to be more important for small than for large diameter beds. Thus in table 8, comparing Pansing's data with those of Mathis and Watson, the bubble diameters are similar with similar bed heights, showing that the wall effect diminishes with larger diameter beds.

The fact that slug flow occurs so frequently with small diameter beds, but is less likely with large diameter beds, should serve as a caution to those who try to scale-up laboratory experiments by means of empirical correlations. For the basic pre-requisite of the dynamical similarity method is that the flow pattern in the model should be geometrically similar to the flow pattern in the full-scale experiment; clearly this condition is not fulfilled if there is slug flow in the model and bubble flow in the full-size reactor. The difficulty can partly be met by applying the theory of this chapter, using values of D_e obtained either from previous experience of full-scale beds, or from expansion data or visual observation of a full-scale bed. It is hoped that further advances in the theory of Chapter 5 will allow D_e, and thence the overall conversion, to be predicted from first principles.

6.5. The importance of heat and mass transfer in a fluidised catalytic reactor

In the above analysis, the only diffusional resistance of any importance is that which governs the mass transfer between the walls of the bubble and the main body of the gas within. It was assumed that there is no diffusional resistance between the gas within the particulate phase and the surface of the particles. This assumption is justified partly by the reasonable results obtained from the above theory, but also more directly from results reported in the literature on gas-to-particle mass transfer in fluidised beds (Chu, 1956; Richardson and Szekely, 1961; Riccetti and Thodos, 1961).

Typical experiments on mass transfer are those in which naphthalene-coated particles are fluidised by air, and the rate of evaporation of naphthalene is measured. Chu (1956) has shown that the results of such experiments can be correlated in much the same way as for a fixed bed of particles. However, in all cases the experiments were conducted on extremely shallow beds of fluidised particles, often only a fraction of an inch high. From the work of Yasui and Johanson (1958) and from what has been reported in earlier chapters, such shallow beds should contain only very small bubbles, so that a mass transfer correlation of the fixed-bed type is to be expected. Now in the experiments with volatile particles, the shallow beds had to be used in order to obtain any results at all; deeper beds gave equilibrium between the particles and the exit gas, and this fact—the rapid establishment of equilibrium—is the real justification for assuming no diffusional resistance within the particulate phase in the reactor system.

Referring to §6.4 (a), p. 119, these mass transfer experiments with very shallow beds are in agreement with the observations of Yasui and Johanson that the first few millimetres above the bed support contain only very small bubbles, and in fact behave more like a particulate than an aggregative bed. Therefore with a very high rate of reaction on the particles, much of the conversion would take place within this very shallow entry region, leading to a better overall conversion than is predicted by (6.9) or (6.19). These equations do not, therefore, apply for infinite values of k', though they are valid for large values of k' such as are shown in fig. 37. The same difficulty makes it very hard to observe by-passing in a mass transfer experiment. For example, with a bed of naphthalene particles fluidised by air, the exit air would not contain the equilibrium proportion of naphthalene if the bubbles were truly uniform right down to the distributor plate; in fact, equilibrium will be established within the first few millimetres, and although bubbles are formed higher up, these bubbles contain air already brought to equilibrium with the particles, and thus the by-passing is not detected. This brings out an extreme subtlety of the fluidised system which is that it can give the appearance of perfect gas–solid contacting with a purely mass transfer system, and yet with a catalysed reaction, such as Orcutt's ozone decomposition, an appreciable proportion of the reactant can pass through the bed unconverted.

Various experiments (Kettenring, Manderfield and Smith 1950; Walton, Olson and Levenspiel, 1952; Wamsley and Johanson, 1954; Heertjes and McKibbins, 1956; and Heertjes, 1962) on heat transfer between the fluidising gas and the particles show a similar picture, that of rapidly established equilibrium within a very short distance above the distributor plate. For example, Heertjes and McKibbins (1956), who studied the drying of air-fluidised silica gel (0·36–1·1 mm) in a shallow fluidised bed, found that an initial temperature difference of 22 °F between the air and the particles had almost disappeared within a height of 6 mm. It therefore seems reasonable to assume that within the particulate phase, the gas and particles are always at the same temperature.

APPENDIX A

A. 1. Irrotational motion past a cylinder or sphere

The problem to be considered is the motion of an inviscid incompressible fluid flowing in a uniform stream with velocity W past a solid cylinder or a solid sphere. The axes, coordinates and velocities are shown in fig. 39. The motion is defined by two physical principles, namely,

(i) the equation of continuity, derived from a material balance on an element fixed in space, and

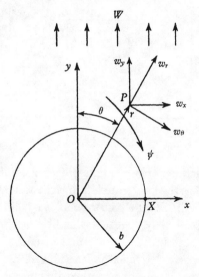

Fig. 39. Coordinates and velocities.

(ii) the fact that since the fluid has zero viscosity, each element must be free from shear stress and therefore cannot acquire angular velocity. The motion is therefore said to be *irrotational*. Also, since there are no shear stresses, Bernoulli's theorem may be used to calculate the pressure at every point.

(a) Continuity in two-dimensional motion

Fig. 40 shows a fixed element $ABCD$, having unit thickness normal to the paper. The flow-rate across AB is $w_x \, dy$ and across CD

is $[w_x + (\partial w_x/\partial x)\,dx]\,dy$, and there are similar terms for the faces AD and BC. Since the fluid is incompressible, the net flow-rate into the element $ABCD$ is zero and therefore

$$\frac{\partial w_x}{\partial x} + \frac{\partial w_y}{\partial y} = 0. \tag{A.1}$$

This equation is satisfied by writing

$$\left.\begin{aligned} w_x &= \frac{\partial \psi}{\partial y}, \\[2mm] w_y &= -\frac{\partial \psi}{\partial x}. \end{aligned}\right\} \tag{A.2}$$

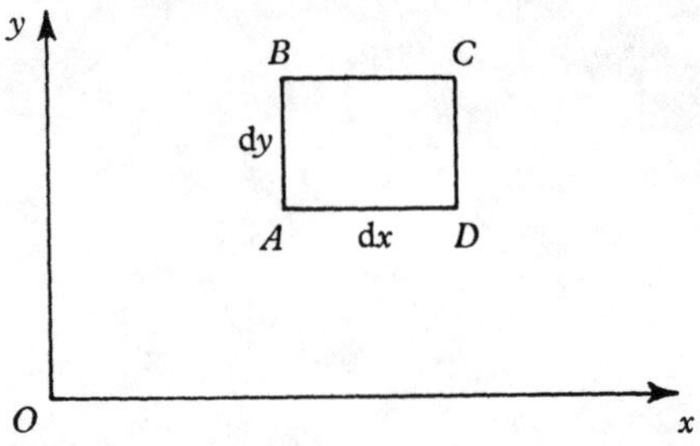

Fig. 40. Rectangular element fixed in space.

The stream function ψ has a simple physical significance in that it can be regarded as the flow from left to right across any line joining the origin O and the point A. Using this as a definition of ψ, (A.2) can be derived directly, because the flow across AB must be $(\partial \psi/\partial y)\,dy = w_x\,dy$, leading to the first of (A.2), and the second equation is similarly derived from the flow across AD. The same process can be carried out in any other direction, so that *the rate of change of ψ in any direction gives the velocity in a perpendicular direction.* If the typical point A is moved about in the (x, y)-plane so that ψ is kept constant, then the curve traced out by A must be a streamline, because there can be no flow across it. For a more complete explanation of the properties of the stream function, the reader should refer to Milne-Thomson (1960, Chapter XV).

(b) Continuity with motion symmetrical about an axis

Referring to fig. 41, the motion is symmetrical about the axis of y, and the material balance is on an element formed by rotating $PQRS$

about the axis of y. The flow-rate across the curved surface formed by PQ is thus $w_\theta 2\pi r \sin\theta\, dr$, and across PR the corresponding quantity is $w_r 2\pi r^2 \sin\theta\, d\theta$. These flow-rates increase differentially across RS and QS, and balancing these differentials to give zero net flow into the element gives

$$\sin\theta\frac{\partial}{\partial r}(r^2 w_r) + r\frac{\partial}{\partial\theta}(w_\theta \sin\theta) = 0. \qquad (A.3)$$

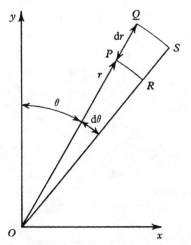

Fig. 41. Element fixed in space and formed by polar coordinates.

This equation is satisfied by writing

$$\left.\begin{aligned}
w_r &= -\frac{1}{r^2\sin\theta}\frac{\partial\psi}{\partial\theta}, \\[2mm]
w_\theta &= \frac{1}{r\sin\theta}\frac{\partial\psi}{\partial r}.
\end{aligned}\right\} \qquad (A.4)$$

The stream function ψ has the same sort of significance as in the two-dimensional case, in that the flow towards the origin through the circle formed by rotating P about the y axis, is $2\pi\psi$. This may be used as a definition of ψ from which (A.4) can be derived. Also, streamlines are curves of constant ψ, as before.

(c) Irrotational motion in two dimensions

It is assumed that the viscosity of the fluid is zero, and hence there can be no shear stress at any point. Each element of fluid is therefore subjected only to normal pressure forces, but since these have no moment about the centre of the element, the latter cannot acquire angular momentum if it has none initially. This is the case in our problem since the fluid begins as a uniform stream and therefore must have zero angular velocity in passing round the cylinder. Expressed quantitatively this means that in fig. 40 the angular velocities of AB and of AD must be equal and opposite, that is,

$$\frac{\partial w_x}{\partial y} = \frac{\partial w_y}{\partial x}. \tag{A.5}$$

This equation is satisfied by writing

$$\left.\begin{array}{l} w_x = \dfrac{\partial \phi}{\partial x}, \\[2mm] w_y = \dfrac{\partial \phi}{\partial y}, \end{array}\right\} \tag{A.6}$$

and this defines the velocity potential ϕ, which has the property, analogous to that of ψ, *that the rate of change of ϕ in any direction is the velocity in that direction.*

Combining (A. 1) and (A. 6) gives Laplace's equation for ϕ,

$$\frac{\partial^2 \phi}{\partial x^2} + \frac{\partial^2 \phi}{\partial y^2} = 0, \tag{A.7}$$

and combining (A. 2) and (A. 5) gives Laplace's equation for ψ,

$$\frac{\partial^2 \psi}{\partial x^2} + \frac{\partial^2 \psi}{\partial y^2} = 0. \tag{A.8}$$

These equations summarise the fact that the conditions of continuity and of no rotation must be satisfied throughout the fluid.

(d) Irrotational motion with axial symmetry

In the same way as for two-dimensional motion, the angular velocity of each element must be zero. In fig. 41, the angular velocity of PR is w_θ/r if the derivatives of w_r and w_θ are zero; when these derivatives are finite, they add angular velocities $\partial w_\theta/\partial r$ to PQ and $-\partial w_r/r\,\partial\theta$ to PR; the sum of these components must be zero, and therefore

$$\frac{\partial}{\partial r}(r w_\theta) - \frac{\partial w_r}{\partial \theta} = 0. \tag{A.9}$$

This equation is satisfied by writing

$$w_r = \frac{\partial \phi}{\partial r}, \\ w_\theta = \frac{1}{r} \frac{\partial \phi}{\partial \theta}, \Bigg\}$$ (A. 10)

the velocity potential ϕ retaining the property that the rate of change of ϕ in any direction gives the velocity in that direction.

Combining (A. 3) and (A. 10) gives the appropriate version of Laplace's equation,

$$\sin\theta \frac{\partial}{\partial r}\left(r^2 \frac{\partial \phi}{\partial r}\right) + \frac{\partial}{\partial \theta}\left(\frac{\partial \phi}{\partial \theta} \sin\theta\right) = 0,$$ (A. 11)

and combining (A. 4) and (A. 9) gives the corresponding equation in ψ,

$$\frac{\partial}{\partial r}\left(\frac{1}{\sin\theta} \frac{\partial \psi}{\partial r}\right) + \frac{\partial}{\partial \theta}\left(\frac{1}{r^2 \sin\theta} \frac{\partial \psi}{\partial \theta}\right) = 0.$$ (A. 12)

(e) Results for flow past a cylinder

The velocity potential and stream function for the flow of a uniform stream past a circular cylinder of radius b are

$$\phi = W\left(r + \frac{b^2}{r}\right)\cos\theta,$$ (A. 13)

$$\psi = -W\left(1 - \frac{b^2}{r^2}\right) r \sin\theta.$$ (A. 14)

By substituting $x = r\sin\theta$ and $y = r\cos\theta$ in (A. 13) and (A. 14) it may be verified that these equations satisfy (A. 7) and (A. 8). The boundary conditions to be satisfied are (i) the velocity should be uniform and equal to W at infinity, and (ii) the radial velocity should be zero at $r = b$. Condition (i) is clearly met by (A. 13) and (A. 14), because when $r \to \infty$, $\phi = Wr\cos\theta = Wy$ and $\partial\phi/\partial y = W$ as it should; also when $r \to \infty$, $\psi = -Wr\sin\theta = -Wx$, and $\partial\psi/\partial x = -W$ as it should. Condition (ii) is also readily verified because when $r = b$, $\partial\phi/\partial r = 0$ from (A. 13), and from (A. 14), when $r = b$, $\psi = 0$, so the surface of the cylinder must be a streamline.

(f) Results for flow past a sphere

The velocity potential and stream function for the flow round a sphere in a uniform stream are

$$\phi = W\left(r + \frac{b^3}{2r^2}\right)\cos\theta, \qquad (A.15)$$

$$\psi = -W\left(1 - \frac{b^3}{r^3}\right)\frac{r^2\sin^2\theta}{2}. \qquad (A.16)$$

It can readily be verified by differentiation that these equations satisfy (A. 11) and (A. 12); that they also satisfy the boundary conditions (i) of uniform velocity W at infinity, and (ii) of zero radial velocity at the surface of the sphere can be verified in the same way as above for the two-dimensional case.

(g) The effective mass M due to the fluid round a moving sphere

The mass M is obtained by calculating the kinetic energy of the fluid set in motion by a sphere moving through fluid which is at rest at infinity. To get the velocity potential for this motion, we add to (A. 15) a term $-Wr\cos\theta$, the velocity potential due to a uniform velocity W, leaving

$$\phi = \frac{Wb^3}{2r^2}\cos\theta. \qquad (A.17)$$

The kinetic energy of this motion is obtained by summing for all elements the quantity $\frac{1}{2}w^2\,dm$, where dm is the mass of the element and w is its velocity. Each element is formed by rotating the area $PQSR$ in fig. 41 about the y axis, giving $dm = \rho 2\pi r^2 \sin\theta\,d\theta\,dr$. Also, $w^2 = (\partial\phi/\partial r)^2 + (\partial\phi/r\,\partial\theta)^2$, and substituting from (A. 17) gives $w^2 = W^2 b^6(1 + 3\cos^2\theta)/4r^6$. The total kinetic energy is therefore

$$\sum \frac{w^2}{2}\,dm = \int_{r=b}^{\infty}\int_{\theta=0}^{\pi} \frac{\pi\rho W^2 b^6}{4r^4}(1 + 3\cos^2\theta)\sin\theta\,d\theta\,dr$$
$$= \tfrac{1}{3}\pi\rho b^3 W^2.$$

To get the effective mass M, this kinetic energy is divided by $\frac{1}{2}W^2$, giving $M = \frac{2}{3}\pi\rho b^3$, which is half the mass of the fluid displaced by the sphere. It is this inertia which has to be overcome when a sphere in a fluid is accelerated, and M is added to the mass of the sphere in calculating the total force necessary to cause the acceleration.

A. 2. Percolation of fluid through a void in a particle bed

This section sets out an analysis of the way in which viscous fluid percolates through a void in a fixed bed of particles, under the influence of a uniform pressure gradient acting throughout the bed. D'Arcy's law is assumed to hold good, so that the fluid velocity is proportional to the pressure gradient. It is then easy to show that the pressure within the bed is governed by Laplace's equation, and this equation is then solved with the boundary conditions of a constant pressure gradient at infinity, and of uniform pressure within the void, which is empty of particles. Solutions are given for a cylindrical void and for a spherical void. From the pressure distribution, D'Arcy's law gives the velocity distribution leading to a stream function from which the streamlines can be plotted.

(a) Derivation of Laplace's equation

The axes and velocities of fig. 39 will be used, the velocities representing *interstitial* values. The voidage ϵ of the bed is taken to be uniform, and the volume flow-rate across unit area within the bed is therefore the velocity normal to the area times the voidage ϵ. The percolating fluid is assumed to be incompressible, and the continuity equations (A. 1) and (A. 3) therefore apply. Now from D'Arcy's law, the velocity in any direction is $-K$ times the pressure gradient in that direction; but a very similar property belongs also to ϕ and therefore (A. 6) and (A. 10) apply to the percolation problem if ϕ is replaced by $-Kp$, where p is the pressure. Combining these equations with the corresponding continuity equations (A. 1) or (A. 3) it follows that Laplace's equation holds for the percolation problem with ϕ replaced by p. Thus (A. 7) applies for two-dimensional percolation, and (A. 11) for axially symmetrical percolation, with p instead of ϕ.

(b) Results for percolation through a cylindrical void

Referring to fig. 39, the boundary conditions to be satisfied are that when $r = b$, p must be constant, and when r is large, dp/dy must be constant $= -W/K$. The latter condition gives a uniform interstitial velocity of W, which is K times the pressure gradient at infinity. The

9

pressure must satisfy (A. 7) throughout the bed, and these conditions are met by the result

$$p = -\frac{W}{K}\left(r - \frac{b^2}{r}\right)\cos\theta. \qquad (A. 18)$$

This has the same form as (A. 13) and therefore clearly satisfies (A. 7); when r is large, $Kp = -Wr\cos\theta$, which represents a uniform pressure gradient and velocity in the y direction; when $r = b$, $p = 0$, thus satisfying the constant pressure requirement at the surface of the void.

A stream function can be calculated for the motion by eliminating the velocities from between (A. 2) and (A. 6), putting $\phi = -Kp$ in the latter; this leads to $\partial\psi/\partial y = -K\,\partial p/\partial x$, $\partial\psi/\partial x = K\,\partial p/\partial y$. These equations, or rather their equivalents in polar coordinates,

$$\partial\psi/\partial r = -K\,\partial p/r\,\partial\theta, \quad \partial\psi/r\,\partial\theta = K\,\partial p/\partial r,$$

can be used to calculate ψ from (A. 18) by integration, giving

$$\psi = -W\left(1 + \frac{b^2}{r^2}\right)r\sin\theta. \qquad (A. 19)$$

Note that ψ gives the flow divided by ϵ, because our analysis is in terms of *interstitial* velocities. Putting $r = b$ and $\theta = \frac{1}{2}\pi$ in (A. 19) gives the flow $\psi\epsilon$ between the origin and point X, fig. 39, which is $-2W\epsilon b$. Ignoring the sign, the total flow through the void is therefore

$$q = 4W\epsilon b. \qquad (A. 20)$$

(c) Results for percolation through a spherical void

The boundary conditions are the same as for the cylindrical void, namely, constant pressure within the void, and a uniform interstitial velocity $-W$ in the y direction at infinity. To satisfy (A. 11), we use an equation of the same form as (A. 15) but satisfying the boundary conditions specified in the previous sentence. The result is

$$p = -\frac{W}{K}\left(r - \frac{b^3}{r^2}\right)\cos\theta. \qquad (A. 21)$$

This can be shown by differentiation to satisfy (A. 11), and to satisfy the boundary conditions in the same way as did (A. 18).

The stream function is calculated as it was for the cylindrical void, by eliminating the velocities from between (A. 4) and (A. 10) with ϕ replaced by $-Kp$. Substitution for p from (A. 21) and integration then gives

$$\psi = -W\left(1 + \frac{2b^3}{r^3}\right)\frac{r^2}{2}\sin^2\theta. \qquad \text{(A. 22)}$$

The flow through the void is the value of $2\pi\psi\epsilon$ at the point X in fig. 39, using the same method as for the cylindrical void, so that in this case

$$q = 3\pi W\epsilon b^2. \qquad \text{(A. 23)}$$

It is interesting to note that the 'short circuiting' effect, whereby the fluid is attracted to the void by its giving an easier path through the bed, leads to a flow-rate of twice the superficial velocity $W\epsilon$ times the exposed area $2b$ of the cylinder, whereas for the sphere, the flow-rate is three times $W\epsilon\pi b^2$, the corresponding flow.

APPENDIX B

The pressure recovery in the wake below a spherical-cap bubble

We wish to calculate the quantity p_R shown in fig. 42 (c); this quantity can be regarded as the pressure recovery between point B in fig. 42 and a point 2, far below the bubble. Thus in the idealised calculations of Rippin (1959), the bubble was assumed to be followed by a stagnant wake, bounded by the broken curves shown in fig. 42 (a). In this case the pressure difference between the horizontal at B and point 2 was merely ρg times the difference in level,

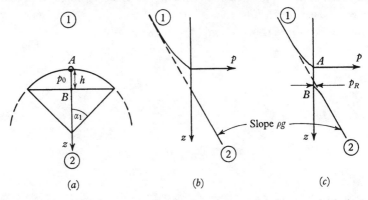

Fig. 42. Sketches of pressure distribution above and below a spherical-cap bubble (a). (b) Due to Rippin (1959). (c) Calculated points A and B from Davies and Taylor's (1950) result.

giving the pressure distribution shown in fig. 42 (b). In an actual bubble, the stagnant wake assumed by Rippin exists only for a very small depth below the bubble; turbulent mixing then leads to the pressure recovery p_R shown in fig. 42 (c).

The quantity p_R can be calculated from the experimental results of Davies and Taylor (1950); they gave the semi-empirical formula for the rising velocity

$$U_b = 0.792 g^{\frac{1}{2}} V^{\frac{1}{6}}, \tag{B.1}$$

V being the bubble volume, and their measured value of α_1, the half angle shown in fig. 42 (a) was 50 degrees. To calculate p_R, we assume the bubble to be held stationary by a downward velocity U_b, and apply Bernoulli's theorem between the stagnation point A and position 1, at height $-z_1$ above A, giving

$$p_0 = p_1 - \rho g z_1 + \tfrac{1}{2}\rho U_b^2, \tag{B.2}$$

z_1 being large and negative, and p_0 the pressure within the bubble. The equation relating the pressures at point A and at position 2 is

$$p_2 = p_0 + \rho g(z_2 - h) + p_R, \tag{B.3}$$

where h is the height of the bubble, and p_R is the pressure recovery indicated in fig. 42 (c). Since the liquid at a great distance from the bubble is moving uniformly, by following a path from 1 to 2 at a great distance from the bubble, we get

$$p_2 = p_1 + \rho g(z_2 - z_1). \tag{B.4}$$

Eliminating p_0, p_1 and p_2 from (B.2), (B.3) and (B.4),

$$p_R = \rho g h - \tfrac{1}{2}\rho U_b^2. \tag{B.5}$$

U_b can be obtained from (B.1); and from the geometry of the bubble (which is assumed to consist of a spherical cap and a flat base with $\alpha_1 = 50°$) we have, from (2.21), $V = 7 \cdot 75 h^3$; substituting this in (B.1) gives $U_b^2 = 1 \cdot 24 g h$, which with (B.5) gives

$$p_R = 0 \cdot 38 \rho g h. \tag{B.6}$$

The pressure distribution is therefore as sketched in fig. 31, p. 78, the points A and B being fixed as shown.

APPENDIX C

Diffusion from the curved surface of a spherical-cap bubble

Referring to fig. 43, it is required to calculate the rate of transfer of a diffusing substance from the surface of the bubble POQ, where the concentration is c^*, to the interior of the bubble, where the concentration is c_0. The fluid mechanics are that the bubble is surrounded by liquid, whose velocity at the surface is $w_s = (2gz)^{\frac{1}{2}}$, and it is assumed that this velocity is imparted to the layer of gas within the bubble and near to the surface POQ. This means that at a given angle α, the velocity at distance y from the point S is assumed to be w_s. In fact, boundary-layer effects will give a variation of velocity with y, so the present analysis will at the best be a first approximation.

To find the equation governing diffusion within the bubble, we consider a balance on the element formed by rotating the infinitesimal area $ABCD$ about the axis Oz, AC and BD being adjacent streamlines. Because these lines are streamlines, there can be no convection across them. Since the film within which diffusion occurs is very thin, diffusion along the streamlines can be neglected in comparison with diffusion across them. In writing down a material balance for the element $ABCD$ there are therefore only two terms, namely:

(a) the difference between diffusion across BD and across AC; this difference must balance

(b) the difference between the convection across AB and across CD.

To derive term (a), we note that the rate of diffusion across AC is $2\pi(R-y)^2 \sin\alpha \, d\alpha \, D_G \, \partial c/\partial y$, since AC is nearly a circular arc. Here c is the concentration of diffusing substance, D_G is its diffusion coefficient, and the other symbols have the significance indicated in fig. 43, the origin of y being at S. The difference between the diffusion across BD and across AC is then

$$2\pi D_G \sin\alpha \, d\alpha \left[R^2 \frac{\partial^2 c}{\partial y^2} \, dy - 2R \frac{\partial c}{\partial y} \, dy \right],$$

where terms in y have been omitted because y is much less than R.

The expression can be further simplified because $\partial c/\partial y$ is much less than $R\,\partial^2 c/\partial y^2$, so that the net diffusion into the element $ABCD$ is

$$2\pi D_G R^2 \sin\alpha \frac{\partial^2 c}{\partial y^2}\,\mathrm{d}y\,\mathrm{d}\alpha. \qquad (\text{C.}1)$$

The convection across AB is $w_s c 2\pi(R-y)\sin\alpha\,\mathrm{d}y$, and if c were constant along the streamlines, the convection across CD would be the same as across AB, because the flow across each line is equal. Hence when there is a gradient of concentration along the streamlines, the net convection out of the element, term (b) above, must be

$$2\pi R w_s \sin\alpha \left(\frac{\partial c}{\partial\alpha}\right)_\psi \mathrm{d}\alpha\,\mathrm{d}y, \qquad (\text{C.}2)$$

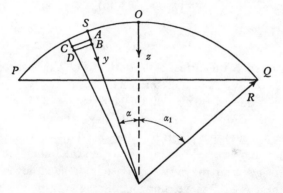

Fig. 43. Coordinates for calculation of diffusion within a bubble.

ψ being the stream function, which must of course be constant along a streamline. Equating (C.1) and (C.2), for a material balance on the element $ABCD$, gives

$$D_G R\left(\frac{\partial^2 c}{\partial y^2}\right)_\alpha = w_s\left(\frac{\partial c}{\partial\alpha}\right)_\psi. \qquad (\text{C.}3)$$

This equation can be put into an integrable form by eliminating y in terms of ψ, the two variables being related by the equation

$$2\pi\psi = 2\pi R w_s y \sin\alpha, \qquad (\text{C.}4)$$

each side of this equation representing the total flow across SA. Now at constant α, w_s does not vary, and therefore ψ is proportional to y, from (C.4), giving $(\partial^2 c/\partial y^2)_\alpha = (\partial^2 c/\partial \psi^2)_\alpha w_s^2 R^2 \sin^2 \alpha$, and substituting this expression in (C.3) gives

$$D_G R^3 w_s \sin^2 \alpha \left(\frac{\partial^2 c}{\partial \psi^2}\right)_\alpha = \left(\frac{\partial c}{\partial \alpha}\right)_\psi. \tag{C.5}$$

This can be converted to the form of the standard diffusion equation by eliminating α in terms of Φ, where Φ is defined by the equation

$$\frac{d\Phi}{d\alpha} = R^3 w_s \sin^2 \alpha. \tag{C.6}$$

Φ is a function of α only, and from (C.5) and (C.6),

$$D_G \left(\frac{\partial^2 c}{\partial \psi^2}\right)_\Phi = \left(\frac{\partial c}{\partial \Phi}\right)_\psi. \tag{C.7}$$

Boundary conditions

(a) Adjacent to the surface of the bubble, POQ, the concentration is assumed to be c^*, and therefore when $\psi = 0$, $c = c^*$.

(b) It is assumed that $c = c_0$ for large values of ψ, and for all values of ψ at $\alpha = 0$. Experiments have shown (Rose, 1961; McWilliam, 1961) that a vortex, rather like Hill's spherical vortex (see Lamb, 1932, p. 246), is formed within the bubble, so that an element of gas near the curved surface is dragged down, and on arriving near P or Q moves horizontally towards the axis Oz, and finally moves up near this axis before beginning the circuit again. By putting $c = c_0$ at $\alpha = 0$, it has therefore been assumed that an element near P or Q, having a concentration greater than c_0, will be fully mixed with the main body of gas before being carried back to O by the vortex. Also, c_0 will increase with time, but it is assumed that these changes are slow compared with the rate at which the concentration profile near SOQ reaches the form given by the present calculations, which assume a constant value of c_0.

With the boundary conditions (a) and (b) above, the solution to (C.7) is

$$\frac{c - c_0}{c^* - c_0} = 1 - \frac{2}{\pi^{\frac{1}{2}}} \int_0^{\psi/2(D_G \Phi)^{\frac{1}{2}}} e^{-\eta^2} d\eta. \tag{C.8}$$

Rate of transfer across the curved surface

Referring to fig. 43, the rate of transfer across the surface POQ is

$$N_c = -D_G \int_0^{\alpha_1} 2\pi R^2 (\partial c/\partial y)_{y=0} \sin\alpha \, d\alpha,$$

α_1 being the maximum value of α. In this equation, y and α can be replaced as coordinates by ψ and Φ, using (C. 4) and (C. 6), giving

$$N_c = -2\pi D_G \int_0^{\Phi_1} \left(\frac{\partial c}{\partial \psi}\right)_{\psi=0} d\Phi, \qquad (C.9)$$

where Φ_1 is obtained from (C. 6) by integration from $\alpha = 0$ to α_1, with the arbitrary assumption that $\Phi = 0$ at $\alpha = 0$. Hence from (C. 6), assuming that $w_s = (2gz)^{\frac{1}{2}}$, where $z = R(1 - \cos\alpha)$,

$$\Phi_1 = R^3 \int_0^{\alpha_1} [2gR(1 - \cos\alpha)]^{\frac{1}{2}} \sin^2\alpha \, d\alpha. \qquad (C.10)$$

Substitution from (C. 8) into (C. 9) gives

$$N_c = 4(\pi D_G \Phi_1)^{\frac{1}{2}} (c^* - c_0). \qquad (C.11)$$

Now an overall mass transfer coefficient, k_G, may be defined in terms of the surface area of a sphere of diameter D_e having the same volume as the bubble. Therefore $N_c = k_G \pi D_e^2 (c^* - c_0)$, and substituting for N_c from (C. 11), and for Φ_1 from (C. 10), k_G is derived, using (2.21) with $\alpha_1 = 50$ degrees, as

$$k_G = 0.975 D_G^{\frac{1}{2}} D_e^{-\frac{1}{2}} g^{\frac{1}{4}}. \qquad (C.12)$$

It is noteworthy that k_G, which has the dimensions of a velocity, does in fact represent the velocity induced by the diffusion. For a gas, D_G is of order 0.1 cm^2/sec, and with $D_e = 1$ cm, k_G is less than 2 cm/sec, which is much less than the rising velocity of a 1 cm bubble, about 30 cm/sec. This justifies the basic assumption that the diffusion boundary layer occupies only a small proportion of the bubble.

The above analysis is similar to that given by Baird and Davidson (1962) for liquid-film controlled diffusion from a rising bubble. They verified experimentally an equation like (C. 12) but with D_G replaced by the liquid-phase diffusion coefficient.

BIBLIOGRAPHY

ADAMS, C. E., GERNAND, M. O. and KIMBERLIN, C. N. (1954). High temperature fluidized solids bath for continuous systems. *Industr. Engng Chem.* **46**, 2458.

ADLER, I. L. and HAPPEL, J. (1962). The fluidization of uniform smooth spheres in liquid media. *Chem. Engng Progr. Symposium Series*, **58**, no. 38, 98.

ANDERSSON, K. E. B. (1961). Pressure drop in ideal fluidization. *Chem. Engng Sci.* **15**, 276.

ASKINS, J. W., HINDS, G. P. and KUNREUTHER, F. (1951). Fluid catalyst gas mixing in commercial equipment. *Chem. Engng Progr.* **47**, 401.

BAIRD, M. H. I. and DAVIDSON, J. F. (1962). Gas absorption by large rising bubbles. *Chem. Engng Sci.* **17**, 87.

BAUMGARTEN, P. K. and PIGFORD, R. L. (1960). Density fluctuations in fluidized beds. *A.I.Ch.E. Journal*, **6**, 115.

BECK, R. A. (1949). Evaluation of fluid catalyst. *Industr. Engng Chem.* **41**, 1242.

BLAKE, F. C. (1922). The resistance of packings to fluid flow. *Trans. Amer. Inst. Chem. Engrs*, **14**, 415.

BLOORE, P. D. and BOTTERILL, J. S. M. (1961). Similarity in behaviour between gas bubbles in liquid and fluidised solid systems. *Nature, Lond.*, **190**, 250.

BROWN, G. G. and associates (1950). *Unit Operations*. New York: Wiley.

CALDERBANK, P. H. (1956). Gas-liquid contacting on plates. *Trans. Instn Chem. Engrs, Lond.*, **34**, 79.

CARMAN, P. C. (1937). Fluid flow through granular beds. *Trans. Instn Chem. Engrs, Lond.*, **15**, 150.

CARMAN, P. C. (1956). *Flow of Gases Through Porous Media*. London: Butterworths.

CHU, J. C. (1956). Heat and mass transfer in solid fluidization (Chapter 2 of *Fluidization*, Edited by D. F. Othmer). New York: Reinhold.

COLLINS, R. E. (1961). *Flow of Fluids Through Porous Materials*. New York: Reinhold.

COULSON, J. M. and RICHARDSON, J. F. (1955). *Chemical Engineering*. London: Pergamon Press.

CRAMPTON, J. R. (1961). The effect of temperature on gas–solids fluidisation. Report for the Chemical Engineering Tripos, University of Cambridge.

DAVIDSON, J. F. (1961). Symposium on fluidisation—Discussion. *Trans. Instn Chem. Engrs, Lond.*, **39**, 230.

DAVIDSON, J. F., PAUL, R. C., SMITH, M. J. S. and DUXBURY, H. A. (1959). The rise of bubbles in a fluidised bed. *Trans. Instn Chem. Engrs, Lond.*, **37**, 323.

DAVIDSON, J. F. and SCHÜLER, B. O. G. (1960). Bubble formation at an orifice in an inviscid liquid. *Trans. Instn Chem. Engrs, Lond.*, **38**, 335.

DAVIDSON, L. and AMICK, E. H. (1956). Formation of gas bubbles at horizontal orifices. *A.I.Ch.E. Journal*, **2**, 337.

DAVIES, R. M. and TAYLOR, Sir GEOFFREY (1950). The mechanics of large bubbles rising through extended liquids and through liquids in tubes. *Proc. Roy. Soc.* A, **200**, 375.

DE KOCK, J. W. (1961). Aggregative fluidisation. Ph.D. dissertation, University of Cambridge.

DIEKMAN, R. and FORSYTHE, W. L. (1953). Laboratory prediction of flow properties of fluidized solids. *Industr. Engng Chem.* **45**, 1174.

DOTSON, J. M. (1959). Factors affecting density transients in a fluidized bed. *A.I.Ch.E. Journal*, **5**, 169.

DUMITRESCU, D. T. (1943). Strömung an einer Luftblase im senkrechten Rohr. *Z. angew. Math. Mech.* **23**, 139.

ERGUN, S. (1952). Fluid flow through packed columns. *Chem. Engng Progr.* **48**, 89.

FURUKAWA, J. and OHMAE, T. (1958). Liquid-like properties of fluidised systems. *Industr. Engng Chem.* **50**, 821.

GARNER, F. H. and HAYCOCK, P. J. (1959). Circulation in liquid drops. *Proc. Roy. Soc.* A, **252**, 457.

GREENSFELDER, B. S., VOGE, H. H. and GOOD, G. M. (1945). Catalytic cracking of pure hydrocarbons. *Industr. Engng Chem.* **37**, 1168.

GRIFFITH, P. and WALLIS, G. B. (1961). Two-phase slug flow. *Trans. Amer. Soc. Mech. Engrs*, **83**, Series C, 307.

GROHSE, E. W. (1955). Analysis of gas-fluidized solid systems by X-ray absorption. *A.I.Ch.E. Journal*, **1**, 358.

HALL, C. C. and CRUMLEY, P. (1952). Some observations on fluidisation as applied to the Fischer–Tropsch process. *J. Appl. Chem.* **2**, S 47.

HAPPEL, J. (1958). Viscous flow in multiparticle systems: slow motion of fluids relative to beds of spherical particles. *A.I.Ch.E. Journal*, **4**, 197.

HAPPEL, J. and BRENNER, H. (1957). Viscous flow in multiparticle systems: motion of spheres and a fluid in a cylindrical tube. *A.I.Ch.E. Journal*, **3**, 506.

HAPPEL, J. and EPSTEIN, N. (1954). Cubic assemblages of uniform spheres. *Industr. Engng Chem.* **46**, 1187.

HARDEBOL, J. (1961). Symposium on fluidisation—Discussion. *Trans. Instn Chem. Engrs, Lond.*, **39**, 229.

HARMATHY, T. Z. (1960). Velocity of large drops and bubbles in media of infinite or restricted extent. *A.I.Ch.E. Journal*, **6**, 281.

HARRISON, D. (1959). Symposium on fluidisation—Discussion. *Trans. Instn Chem. Engrs, Lond.*, **37**, 328.

HARRISON, D., DAVIDSON, J. F. and DE KOCK, J. W. (1961). On the nature of aggregative and particulate fluidisation. *Trans. Instn Chem. Engrs, Lond.*, **39**, 202.

HARRISON, D. and LEUNG, L. S. (1961). Bubble formation at an orifice in a fluidised bed. *Trans. Instn Chem. Engrs, Lond.*, **39**, 409.

HARRISON, D. and LEUNG, L. S. (1962a). The rate of rise of bubbles in fluidised beds. *Trans. Instn Chem. Engrs, Lond.*, **40**, 146.

HARRISON, D. and LEUNG, L. S. (1962b). *Symposium on the Interaction between Fluids and Particles*, p. 127. London: Instn Chem. Engrs.

HASSETT, N. J. (1963). Developments in applications of fluidisation. *Chem. Process Engng*, **44**, 127.

HAWKSLEY, P. G. W. (1951). Some aspects of fluid flow. The effect of concentration on the settling of suspensions and flow through porous media. *Papers Presented at a Conference Organised by the Institute of Physics*. London: Edward Arnold.

HAWTHORN, E., SHORTIS, L. P. and LLOYD, J. E. (1960). The fluidised solids dryway process for the production of uranium tetrafluoride at Springfields. *Trans. Instn Chem. Engrs, Lond.*, **38**, 197.

HEERTJES, P. M. (1962). Simultaneous heat and mass transfer in a fluidized bed of drying silica gel. *Canad. J. Chem. Engng*, **40**, 105.

HEERTJES, P. M. and McKIBBINS, S. W. (1956). The partial coefficient of heat transfer in a drying fluidized bed. *Chem. Engng Sci.* **5**, 161.

HEYWOOD, H. (1962). *Symposium on the Interaction between Fluids and Particles*, p. 1. London: Instn Chem. Engrs.

JACKSON, R. (1963). The mechanics of fluidised beds. Part 1. The stability of the state of uniform fluidisation. Part 2. The motion of fully developed bubbles. *Trans. Instn Chem. Engrs, Lond.*, **41**, 13.

JAEGER, C. (1956). *Engineering Fluid Mechanics*. London: Blackie.

JOBES, C. W. (1954). Fluidized crystal drier pays off. *Chem. Engng*, **61**, Jan., 66.

KETTENRING, K. N., MANDERFIELD, E. L. and SMITH, J. M. (1950). Heat and mass transfer in fluidized systems. *Chem. Engng Progr.* **46**, 139.

KLEE, A. J. and TREYBAL, R. E. (1956). Rate of rise or fall of liquid drops. *A.I.Ch.E. Journal*, **2**, 444.

KOZENY, J. (1927). Über kapillare Leitung des Wassers im Boden. *S.B. Akad. Wiss. Wien*, Abt. II a, **136**, 271.

KRAMERS, H. (1951). On the 'viscosity' of a bed of fluidized solids. *Chem. Engng Sci.* **1**, 35.

LAMB, Sir HORACE (1932). *Hydrodynamics*, 6th edition. Cambridge.

LANNEAU, K. P. (1960). Gas–solids contacting in fluidized beds. *Trans. Instn Chem. Engrs, Lond.*, **38**, 125.

LEUNG, L. S. (1961). Bubbles in fluidised beds. Ph.D. dissertation, University of Cambridge.

LEVA, M. (1959). *Fluidization*. New York: McGraw Hill.

LEWIS, W. K., GILLILAND, E. R. and BAUER, W. C. (1949). Characteristics of fluidized particles. *Industr. Engng Chem.* **41**, 1104.

LEWIS, W. K., GILLILAND, E. R. and GLASS, W. (1959). Solid-catalyzed reaction in a fluidized bed. *A.I.Ch.E. Journal*, **5**, 419.

LINDSAY, R. K. (1961). The effect of temperature on gas–solids fluidisation. Report for the Chemical Engineering Tripos, University of Cambridge.

LOEFFLER, A. L. and RUTH, B. F. (1959). Particulate fluidization and sedimentation of spheres. *A.I.Ch.E. Journal*, **5**, 310.

MARTIN, J. J., McCABE, W. L. and MONRAD, C. C. (1951). Pressure drop through stacked spheres: effect of orientation. *Chem. Engng Progr.* **47**, 91.

MASSIMILLA, L. and JOHNSTONE, H. F. (1961). Reaction kinetics in fluidized beds. *Chem. Engng Sci.* **16**, 105.

MASSIMILLA, L. and WESTWATER, J. W. (1960). Photographic study of solid–gas fluidization. *A.I.Ch.E. Journal*, 6, 134.

MATHESON, G. L., HERBST, W. A. and HOLT, P. H. (1949). Characteristics of fluid–solid systems. *Industr. Engng Chem.* 41, 1099.

MATHIS, J. F. and WATSON, C. C. (1956). Effect of fluidization on catalytic cumene dealkylation. *A.I.Ch.E. Journal*, 2, 518.

MAY, W. G. (1959). Fluidized-bed reactor studies. *Chem. Engng Progr.* 55, Dec., 49.

MCWILLIAM, J. A. C. (1961). The circulation of air within a simulated fluidised bed bubble. Report for the Chemical Engineering Tripos, University of Cambridge.

MILNE-THOMSON, L. M. (1960). *Theoretical Hydrodynamics*, 4th edition. London: Macmillan.

MORSE, R. D. and BALLOU, C. O. (1951). The uniformity of fluidization—its measurement and use. *Chem. Engng Progr.* 47, 199.

MURPHREE, E. V., BROWN, C. L., FISCHER, H. G. M., GOHR, E. J. and SWEENEY, W. J. (1943). Fluid catalyst process. *Industr. Engng Chem.* 35, 768.

NICKLIN, D. J. (1961). Two-phase flow in vertical tubes. Ph.D. dissertation, University of Cambridge.

NICKLIN, D. J. (1962). Two-phase bubble flow. *Chem. Engng Sci.* 17, 693.

NICKLIN, D. J., WILKES, J. O. and DAVIDSON, J. F. (1962). Two-phase flow in vertical tubes. *Trans. Instn Chem. Engrs, Lond.*, 40, 61.

ORCUTT, J. C. (1960). Solids–gas contacting in fluidized beds. Ph.D. dissertation, University of Delaware.

ORCUTT, J. C., DAVIDSON, J. F. and PIGFORD, R. L. (1962). Reaction time distributions in fluidized catalytic reactors. *Chem. Engng Progr. Symposium Series*, 58, no. 38, 1.

OTHMER, D. F. (1956). *Fluidization*. New York: Reinhold.

PANSING, W. F. (1956). Regeneration of fluidized cracking catalysts. *A.I.Ch.E. Journal*, 2, 71.

PIGFORD, R. L. (1959). Hydrodynamic stability of a fluidized bed. Private communication.

PINCHBECK, P. H. and POPPER, F. (1956). Critical and terminal velocities in fluidisation. *Chem. Engng Sci.* 6, 57.

PRANDTL, L. (1952). *The Essentials of Fluid Dynamics*. London: Blackie.

PYLE, D. L. and ROSE, P. L. (1962). *Unpublished Calculations*. Cambridge.

QUIGLEY, C. J., JOHNSON, A. I. and HARRIS, B. L. (1955). Size and mass transfer studies of gas bubbles. *Chem. Engng Progr. Symposium Series*, 51, no. 16, 31.

REID, R. C. and SHERWOOD, T. K. (1958). *The Properties of Gases and Liquids*. New York: McGraw Hill.

REUTER, H. (1963a). Druckverteilung um Blasen im Gas–Feststoff-Fließbett. *Chem. Ing. Techn.* 35, 98.

REUTER, H. (1963b). Mechanismus der Blasen im Gas–Feststoff-Fließbett. *Chem. Ing. Techn.* 35, 219.

RICCETTI, R. E. and THODOS, G. (1961). Mass transfer in the flow of gases through fluidized beds. *A.I.Ch.E. Journal*, 7, 442.

RICE, W. J. and WILHELM, R. H. (1958). Surface dynamics of fluidized beds and quality of fluidization. *A.I.Ch.E. Journal*, 4, 423.

RICHARDSON, J. F. and MEIKLE, R. A. (1961). Sedimentation and fluidisation. Part III. The sedimentation of uniform fine particles and of two-component mixtures of solids. *Trans. Instn Chem. Engrs, Lond.*, 39, 348.

RICHARDSON, J. F. and SZEKELY, J. (1961). Mass transfer in a fluidised bed. *Trans. Instn Chem. Engrs, Lond.*, 39, 212.

RICHARDSON, J. F. and ZAKI, W. N. (1954). Sedimentation and fluidisation. Part I. *Trans. Instn Chem. Engrs, Lond.*, 32, 35.

RIPPIN, D. W. T. (1959). The rise of gas bubbles in liquids. Ph.D. dissertation, University of Cambridge.

ROMERO, J. B. and JOHANSON, L. N. (1962). Factors affecting fluidized bed quality. *Chem. Engng Progr. Symposium Series*, 58, no. 38, 28.

ROSE, P. L. (1961). Circulation in a spherical-cap bubble. Report for the Chemical Engineering Tripos, University of Cambridge.

ROWE, P. N. (1961). Drag forces in a hydraulic model of a fluidised bed, Part II. *Trans. Instn Chem. Engrs, Lond.*, 39, 175.

ROWE, P. N. (1962 a). The effect of bubbles on gas–solids contacting in fluidised beds. *Chem. Engng Progr. Symposium Series*, 58, no. 38, 42.

ROWE, P. N. (1962 b). Private communication. Atomic Energy Research Establishment, Harwell.

ROWE, P. N. and HENWOOD, G. A. (1961). Drag forces in a hydraulic model of a fluidised bed. Part I. *Trans. Instn Chem. Engrs, Lond.*, 39, 43.

ROWE, P. N. and PARTRIDGE, B. A. (1962). *Symposium on the Interaction between Fluids and Particles*, p. 135. London: Instn Chem. Engrs.

ROWE, P. N., PARTRIDGE, B. A., LYALL, E. and ARDRAN, G. M. (1962). Bubbles in fluidized beds. *Nature, Lond.*, 195, 278.

ROWE, P. N. and STAPLETON, W. M. (1961). The behaviour of 12 inch diameter gas-fluidised beds. *Trans. Instn Chem. Engrs, Lond.*, 39, 181.

ROWE, P. N. and WACE, P. F. (1960). Gas flow patterns in fluidised beds. *Nature, Lond.*, 188, 737.

RUBEY, W. W. (1933). Settling velocities of gravel, sand, and silt particles. *Amer. J. Sci.* 25, 325.

SCHEIDEGGER, A. E. (1957). *The Physics of Flow Through Porous Media.* University of Toronto Press.

SCHÜGERL, K., MERZ, M. and FETTING, F. (1961). Rheologische Eigenshaften von gasdurchströmten Fliessbettsystemen. *Chem. Engng Sci.* 15, 1.

SCHÜLER, B. O. G. (1959). Bubble formation at an orifice. Ph.D. dissertation, University of Cambridge.

SHEN, C. Y. and JOHNSTONE, H. F. (1955). Gas–solids contact in fluidized beds. *A.I.Ch.E. Journal*, 1, 349.

SIMPSON, H. C. and RODGER, B. W. (1962). The fluidization of light solids by gases under pressure and heavy solids by water. *Chem. Engng Sci.* 16, 179.

SQUIRES, A. M. (1962). Species of fluidization. *Chem. Engng Progr.* 58, April, 66.

SZEKELY, J. (1962). *Symposium on the Interaction between Fluids and Particles*, p. 197. London: Instn Chem. Engrs.

THOMPSON, R. B. and MACASKILL, D. (1955). Recovery of sulphur from low-grade ore. *Chem. Engng Progr.* **51**, 369.

TIETJENS, O. G. (1934). *Applied Hydro- and Aeromechanics*. New York: McGraw Hill. Reprinted Dover, New York (1957).

TOOMEY, R. D. and JOHNSTONE, H. F. (1952). Gaseous fluidization of solid particles. *Chem. Engng Progr.* **48**, 220.

UNO, S. and KINTNER, R. C. (1956). Effect of wall proximity on the rate of rise of single air bubbles in a quiescent liquid. *A.I.Ch.E. Journal*, **2**, 420.

VAN DEEMTER, J. J. (1961). Mixing and contacting in gas–solid fluidized beds. *Chem. Engng Sci.* **13**, 143.

VAN KREVELEN, D. W. and HOFTIJZER, P. J. (1950). Studies of gas bubble formation. Calculation of interfacial area in bubble contactors. *Chem. Engng Progr.* **46**, 29.

VOLK, W., JOHNSON, C. A. and STOTLER, H. H. (1962). Effect of reactor internals on quality of fluidization. *Chem. Engng Progr.* **58**, March, 44.

WACE, P. F. and BURNETT, S. J. (1961). Flow patterns in gas fluidised beds. *Trans. Instn Chem. Engrs, Lond.*, **39**, 168.

WALLIS, G. B. (1962). Two-phase flow aspects of pool boiling from a horizontal surface. *Instn Mech. Engrs*. Symposium on two-phase fluid flow, London.

WALTERS, J. K. (1962). The dynamics of the initial motion of a bubble and its application to leakage from sieve trays. Ph.D. dissertation, University of Cambridge.

WALTERS, J. K. and DAVIDSON, J. F. (1962). The initial motion of a gas bubble formed in an inviscid liquid. Part 1. The two-dimensional bubble. *J. Fluid Mech.* **12**, 408.

WALTERS, J. K. and DAVIDSON, J. F. (1963). The initial motion of a gas bubble formed in an inviscid liquid. Part 2. The three-dimensional bubble. *J. Fluid Mech.* **17**.

WALTON, J. S., OLSON, R. L. and LEVENSPIEL, O. (1952). Gas–solid film coefficients of heat transfer in fluidized coal beds. *Industr. Engng Chem.* **44**, 1474.

WAMSLEY, W. W. and JOHANSON, L. N. (1954). Fluidised bed heat transfer. *Chem. Engng Progr.* **50**, 347.

WILHELM, R. H. and KWAUK, M. (1948). Fluidization of solid particles. *Chem. Engng Progr.* **44**, 201.

YASUI, G. (1956). Characteristics of air pockets in fluidized beds. Ph.D. dissertation, University of Washington.

YASUI, G. and JOHANSON, L. N. (1958). Characteristics of gas pockets in fluidized beds. *A.I.Ch.E. Journal*, **4**, 445.

ZENZ, F. A. (1957*a*). *Encyclopedia of Chemical Technology*. First supplement volume. Ed. Kirk, R. E. and Othmer, D. F. Article on fluidization, pp. 365–401. New York: Interscience.

ZENZ, F. A. (1957*b*). Contact efficiency influences design. *Petrol. Refin.* **36**, Nov., 321.

ZENZ, F. A. and OTHMER, D. F. (1960). *Fluidization and Fluid-Particle Systems*. New York: Reinhold.

INDEX OF AUTHORS

Adams, C. E., 1
Adler, I. L., 16, 18
Amick, E. H., 53, 56
Andersson, K. E. B., 12
Ardran, G. M., 30
Askins, J. W., 103, 108

Baird, M. H. I., 137
Ballou, C. O., 19
Bauer, W. C., 16
Baumgarten, P. K., 19, 30, 42
Beck, R. A., 7
Blake, F. C., 12
Bloore, P. D., 59, 60, 61
Botterill, J. S. M., 59, 60, 61
Brenner, H., 18
Brown, C. L., 2
Brown, G. G., 12
Burnett, S. J., 64, 71

Calderbank, P. H., 53
Carman, P. C., 9, 11, 12, 13
Chu, J. C., 121
Collins, R. E., 12
Coulson, J. M., 12
Crampton, J. R., 92
Crumley, P., 2, 7

Davidson, J. F., 22, 29, 30, 31, 32, 33, 34, 36, 54, 55, 56, 65, 73, 80, 137
Davidson, L., 53, 56
Davies, R. M., 22, 24, 25, 26, 32, 33, 34, 35, 37, 38, 47, 66, 85, 132
de Kock, J. W., 37, 73, 80, 81, 82
Diekman, R., 40
Dotson, J. M., 19
Dumitrescu, D. J., 22, 36
Duxbury, H. A., 30

Epstein, N., 16
Ergun, S., 12

Fetting, F., 40
Fischer, H. G. M., 2
Forsythe, W. L., 40
Furukawa, J., 40

Garner, F. H., 83
Gernand, M. O., 1
Gilliland, E. R., 16, 97

Glass, W., 97
Gohr, E. J., 2
Good, G. M., 110
Greensfelder, B. S., 110
Griffith, P., 27
Grohse, E. W., 7

Hall, C. C., 2, 7
Happel, J., 16, 17, 18
Hardebol, J., 92
Harmathy, T. Z., 26
Harris, B. L., 56
Harrison, D., 30, 31, 32, 33, 34, 36, 39, 40, 41, 43, 45, 46, 47, 48, 49, 50, 55, 57, 58, 59, 60, 80, 81, 85
Hassett, N. J., 8
Hawksley, P. G. W., 18
Hawthorn, E., 2
Haycock, P. J., 83
Heertjes, P. M., 122
Henwood, G. A., 15
Herbst, W. A., 40
Heywood, H., 85
Hinds, G. P., 103
Hoftijzer, P. J., 53
Holt, P. H., 40

Jackson, R., 63, 82
Jaeger, C., 65
Jobes, C. W., 2
Johanson, L. N., 19, 30, 35, 42, 62, 88, 89, 110, 119, 121, 122
Johnson, A. I., 56
Johnson, C. A., 7
Johnstone, H. F., 19, 97, 99, 114, 115

Kettenring, K. N., 122
Kimberlin, C. N., 1
Kintner, R. C., 31, 32, 33, 35, 36
Klee, A. J., 26, 27
Kozeny, J., 9, 18
Kramers, H., 40
Kunreuther, F., 103
Kwauk, M., 4, 5, 16, 90

Lamb, Sir Horace, 136
Lanneau, K. P., 19, 41, 42
Leung, L. S., 5, 30, 31, 32, 33, 34, 36, 39, 40, 41, 43, 45, 46, 47, 48, 49, 50, 56, 57, 58, 59, 60, 80, 89

Leva, M., 7, 13, 14
Levenspiel, O., 122
Lewis, W. K., 16, 97, 99, 109, 113, 114
Lindsay, R. K., 92
Lloyd, J. E., 2
Loeffler, A. L., 18
Lyall, E., 30

MacAskill, D., 2
McCabe, W. L., 15
McKibbins, S. W., 122
McWilliam, J. A. C., 83, 93, 136
Manderfield, E. L., 122
Martin, J. J., 15
Massimilla, L., 7, 97, 99, 114, 115
Matheson, G. L., 40
Mathis, J. F., 97, 99, 110, 111, 115, 120
May, W. G., 97, 99, 104, 113, 115, 116, 117
Meikle, R. A., 18
Merz, M., 40
Milne-Thomson, L. M., 56, 124
Monrad, C. C., 15
Morse, R. D., 19
Murphree, E. V., 2

Nicklin, D. J., 22, 27, 28, 29, 41

Ohmae, T., 40
Olson, R. L., 122
Orcutt, J. C., 29, 98, 99, 103, 107, 108, 109, 113, 116
Othmer, D. F., 2, 30, 69, 98

Pansing, W. F., 97, 99, 112, 113, 120
Partridge, B. A., 30, 54
Paul, R. C., 30
Pigford, R. L., 19, 29, 30, 42, 82
Pinchbeck, P. H., 90
Popper, F., 90
Prandtl, L., 9, 11, 12
Pyle, D. L., 94, 95, 96

Quigley, C. J., 56

Reid, R. C., 107, 111
Reuter, H., 77
Riccetti, E. R., 121
Rice, W. J., 62, 81
Richardson, J. F., 12, 16, 17, 18, 121
Rippin, D. W. T., 23, 24, 26, 132
Rodger, B. W., 80, 87
Romero, J. B., 19, 88

Rose, P. L., 83, 93, 94, 95, 96, 136
Rowe, P. N. 7, 14, 15, 30, 31, 39, 54, 71, 74, 75, 94
Rubey, W. W., 85
Ruth, B. F., 18

Scheidegger, A. E., 12
Schügerl, K., 40, 41
Schüler, B. O. G., 53, 55, 56
Shen, C. Y., 97, 99, 115
Sherwood, T. K., 107, 111
Shortis, L. P., 2
Simpson, H. C., 80, 87
Smith, J. M., 122
Smith, M. J. S., 30
Squires, A. M., 8
Stapleton, W. M., 7
Stotler, H. H., 7
Sweeney, W. J., 2
Szekely, J., 117, 118, 121

Taylor, Sir Geoffrey, 22, 24, 25, 26, 32, 33, 34, 35, 37, 38, 47, 66, 85, 132
Thodos, G., 121
Thompson, R. B., 2
Tietjens, O. G., 54
Toomey, R. D., 19
Treybal, R. E., 26, 27

Uno, S., 31, 32, 33, 35, 36

van Deemter, J. J., 97, 99, 104
van Krevelen, D. W., 53
Voge, H. H., 110
Volk, W., 7

Wace, P. F., 64, 71
Wallis, G. B., 27, 29
Walters, J. K., 53, 54, 55, 58
Walton, J. S., 122
Wamsley, W. W., 122
Watson, C. C., 97, 99, 110, 111, 115, 120
Westwater, J. W., 7
Wilhelm, R. H., 4, 5, 16, 62, 81, 90
Wilkes, J. O., 22

Yasui, G., 19, 30, 35, 42, 62, 88, 110, 119, 121, 122

Zaki, W. N., 16, 17, 18
Zenz, F. A., 2, 13, 30, 69, 98

GENERAL INDEX

absolute velocity, *see* interstitial velocity
acceleration of sphere in fluid, 128
  see also bubble acceleration
adsorption, 117–18
agglomeration, 2, 7, 89
aggregative fluidisation, 3, 5, 80, 82
  expansion of aggregative bed, *see* expansion
  effect of fine particles, 88
  shallow beds, 122
  transition to particulate fluidisation, 80–1, 86, 89–91, 94
  two-phase theory, 19–20, 42
air bubble
  in fluidised bed, 3, 21, 30–1, 74–5, 82, 89
  in liquid, 24, 27, 30, 40, 50, 53, 58
air fluidised bed, 3, 4, 21
  bubble: coalescence, 49; formation, 57–8; rising, 30–1, 74–5, 110; stability, 80, 82, 85–6, 89, 92
  mass and heat transfer, 122
  reaction, 107, 112, 115–17
  slug flow, 41
alumina, 99, 110, 113–14, 117
ammonia, 99, 107, 114
analogy with gas-liquid, 35, 41, 46, 50, 62–3, 82; *see also* inviscid liquid, and two-phase
anemometer, 82–3
angular velocity of element, 123, 126
aqueous glycerol, *see* glycerol
arcton, 80
argon, 80
Arnold's method for estimating D_G, 107, 111
asymptotic conversion, *see* conversion
axi-symmetric motion, 68–9, 124–7, 129–31

baffles, effect on fluidisation, 7–8
Ballotini, 30, 60–1, 71
bed depth, *see* height
bed diameter, *see* diameter
bed expansion, *see* expansion, and height
bed height, *see* height
benzene, 107, 111
Bernoulli's theorem, 22–3, 26, 66, 75, 123, 133

boundary conditions
  bubble in: fluidised bed, 65, 67, 75; large volume of liquid, 24–5; tube, 22
  diffusion within bubble, 136
  flow round: cylinder, 127; sphere, 128
  fluidised reactor, 101, 105
  fluidising fluid, 65
  percolation, 129–30
boundary layer, 134, 137
bubble
  acceleration, 52, 54
  air, *see* air bubble
  boundary conditions, *see* boundary
  break up, 37, 82, 92,
  circulation within, *see* circulation
  cloud, *see* bubbles
  coalescence, *see* coalescence
  concentration within, *see* concentration
  density, *see* density
  diffusion from, *see* diffusion
  distortion, 54
  double formation, *see* double
  effective mass, *see* inertia
  elongation, 49, 82
  entrance effect, *see* entrance
  eruption, *see* bubble surfacing
  expansion on rising, 39, 93
  flow, *see* incipient fluidisation
  flow to and from, *see* exchange
  frequency, *see* frequency
  gas, *see* gas
  gas film resistance, *see* diffusion
  height, 22–3, 110, 114, 133
  injected, *see* injected
  long, *see* bubble, rising
  material balance, *see* material balance
  nose, *see* nose
  phase, 19, 42, 63, 98, 102, 108, 112–14; *see also* exchange
  radius of curvature, 24–5
  Reynolds number, 41
  roof, stability, 63, 77, 79
  shape, 71; *see also* bubble, spherical-cap, angle
  size, *see* bubble diameter
  sizes in a fluidised bed, 62
  spherical, 76, 78
  surfacing, 3, 29, 33, 38–9, 43, 73–4, 92–3

bubble (*cont.*)
thickness, *see* thickness
through-flow, *see* exchange
three-dimensional, 39, 54, 67, 69, 72
time of: coalescence, 44–5; formation 52; rise, 31, 33, 36
transfer coefficient within, 101, 106, 111–12, 119, 137; *see also* mass transfer
two-dimensional, 30–1, 39, 43, 54, 64, 69–70, 79
water, 37, 73–4, 82, 89
bubbles
account for excess flow, *see* incipient fluidisation
buffeting due to, 3
circulation due to, *see* circulation
cloud of, 28, 100, 120
continuously generated, *see* continuously
excess flow due to, *see* incipient fluidisation
expansion due to, *see* expansion
in liquids, 21–9
interaction, 42–9, 55–60, 120
maximum size, *see* bubble stability, and bubble diameter
origin in fluidised beds, 81
relative velocity between, 45–7
role in two-phase theory, *see* analogy
stream, 21, 26; in a tube, 26–8; in a large tube, 28–9
velocity in slug flow, 41–2
water, *see* bubble, water
bubble area, 101, 106, 111, 113–14, 137
bubble-cap, plate or tray, 7, 51, 61, 97
bubble diameter
effect of bed height, *see* height
equivalent, 24, 26, 29, 32, 40, 77, 82, 85–93, 109; from reaction data, 98–100, 105–6, 109–21, 137
frontal, 5, 32, 38–9, 46–7, 92
large equipment, 116–17
maximum, 49, 84–93, 97–8, 117
bubble formation, 50–62
adsorption during, 117–18
at a distributor, 49, 61–2
distortion during, 54
double, *see* double
effects of: orifice, 55–6; varying airflow, 56
experimental results: liquids, 53–6; fluidised beds, 57–61

from small disturbance, 82
in: fluidised bed, 50, 57–62; inviscid liquid, 50–6
irregular, 58
leakage during, 36, 60–1; *see also* exchange
multiple, 58, *see also* double bubble formation
parallel orifices, 61–2
quadruple, *see* quadruple
bubble, rising
fluid flow within, *see* circulation
in: a tube, 21–3, 31–2, 35; an infinite liquid, 22–6; a quiescent bed, 31–4
material balance on, 100, 104
motion of: particles, 66–74; fluidising fluid, 66–74, 94–6
particle and fluid streamlines due to, 66–7, 68–74, 94–6
pressure distribution around, *see* pressure distribution
bubble, spherical-cap
angle subtended or included, 30, 38–41, 92, 132–3, 137; *see also* shape
in: liquid, 23–6, 43, 47, 54, 66, 132; fluidised bed, 31, 48, 66, 71, 74, 76–8, 82–4, 93, 96; *see also* bubble wake, circulation, diffusion, and water bell
bubble stability
effects of: density ratio, 87–8; particle size, 88–9; viscosity of fluidising fluid, 89
experiments, 87–93
fluidised beds, 77–96
gas and liquid fluidised beds, 84, 89
governs maximum size, 84–7, 91–3
theory, 93–6
bubble velocity, 22, 24–5, 29–42, 48, 77, 85, 99–100, 116, 120, 132, 137
compared with incipient fluidising velocity, 71–4, 94–6
effects of: bed viscosity, 40–1, 78; particles, 31, 34–5; shape, 38–40
initial velocity, 31
related to circulation within, 82–4; *see also* circulation, and velocity
relative, between bubbles, 45–7
bubble volume, 24, 31, 34, 36–9, 45, 48, 84–5, 132
at an orifice, 48, 51–5, 59–62; *see also* bubble diameter, and injected bubble

bubble wake, 22–5, 43–9, 54, 71, 74
  pressure recovery in, 132
  stability, 82, 84, 93, 96
bulk flow from bubble, *see* exchange
buoyancy, 50–2, 60
by-passing, 3, 63, 103, 108, 117–18, 122; *see also* exchange

capacitance
  meter, 56
  probe, 19, 41, 57–8
carbon dioxide, fluidising: micro balloons, 80, 89–90; resin particles, 5
carbon on catalyst, 112
carbon tetrachloride, 107
  injection, 117–18
catalyst, 41, 74, 89, 92, 103, 109–10, 112–14, 117
  regenerator, 108, 112–13
  spent, 112
catalytic
  adsorption of carbon tetrachloride, 117–18
  conversion: experiments, 99, 106–15; theory, 98–106, 119
  converters, 119; heat and mass transfer, 121–2
  cracking, 112–13
  dealkylation of cumene, 110–12
  decomposition of nitrous oxide, 115
  hydrogenation of ethylene, 113–14
  oxidation of ammonia, 114–15
  reactions, 2, 20; fast, 103; fluidised bed, 97–122
  reactor, 69
cavity, *see* void
chain of bubbles, *see* bubbles, stream
channelling, 89
  and spouted beds, 6–7, 20, 37
ciné photography, *see* photography
circle of penetration, *see* penetration
circulation
  current due to bubbles, 54
  fluid in bubble, 82–4, 93–6; *see also* velocity within bubble
  liquid in drop, 26, 82–3
  particles in fluidised bed, 18
cloud of bubbles, *see* bubbles, cloud of
coalescence of bubbles, 7, 21, 42–9
  at an orifice, 55
  experiments, 43–7
  in a fluidised bed, 62, 91, 97, 120
  time of, 44–5
  two-phase hypothesis, 47–8

coefficient, *see* transfer, and diffusion
column diameter, *see* diameter
combustion in a catalyst regenerator, 99, 112–13
complete mixing in the particulate phase, 100–3, 105, 113; *see also* mixing
computer, 24, 116
concentration of reactant, 101
  difference between phases, 112
  entering and leaving bed, 100–5, 117
  particulate phase, 100–5, 108
  within bubble, 100–5, 108, 117, 134–7
constant temperature bath, 1
contacting, gas-particle, 119, 122
continuity equation, 29, 123–4, 129
  fluidising fluid, 65, 100
  particles, 64
continuity of operation, 2
continuous phase, 26, 83; *see also* particulate phase
continuously generated bubbles, 26–9, 41–2
  velocity in: fluidised bed, 41–2; water, 26–9; *see also* bubbles, cloud; and bubbles, stream
convection within bubble, 134–5
conversion
  asymptotic, 103, 107–9, 122
  effects of height, velocity, diameter, 108
  fast reaction, 103, 108, 119, 122
  in terms of bubble diameter, 105–6
  near bed support, 119, 122
  overall, 97, 102–3, 105, 119, 121–2
  perfect mixing, 100–3
  piston flow, 103–5
  see also catalytic conversion
converters, *see* catalytic converters
cracking, *see* catalytic cracking
cross-flow ratio, 116; *see also* exchange
cubic mode of packing, 13
cumene, 99, 107, 110–12
curtains of particles, 79
curvature, *see* bubble, radius of curvature
cylinder, irrotational motion past, 123, 127
cylindrical void, percolation, 69, 129–30; *see also* exchange

D'Arcy's law, 64–5, 129
dealkylation, *see* cumene
degree of conversion, *see* conversion

density
  bubble phase, 37, 85
  continuous phase, 26, 81
  difference, solid-fluid, 85–6, 88
  drops, 26
  fluidising fluid, 90–1
  liquid, 52
  particles, 85–7, 90–1, 110, 117–18
  particulate phase, 37, 41, 75
  ratio, particles to fluid, 81
dense phase, see particulate phase
depth of bed, see height
diameter of bed, column, vessel or
    tube compared with bubble dia-
    meter, 22, 31–2, 35–8, 41, 46–7,
    82, 92, 99, 109–11, 113–14, 116,
    118, 120
  effect on conversion, 108–9
  see also bubble diameter, hydraulic
    mean diameter, particle diameter,
    tube diameter
diffusion
  coefficient, gas-phase, 107, 109, 111,
    113–16, 118, 134, 137
  from bubbles, 97, 99, 106, 120–1
  in the particulate phase, 99, 104,
    121
  velocity, 137
  within bubble, 134–7
dilute-phase fluidisation, 6, 16
disadvantages of fluidised systems, 2
distributor, 7, 49, 61–2, 97, 122
double-bubble formation, 55, 58
drag coefficient
  isolated sphere, 12–13, 15
  sphere in a packed bed, 14
drag forces
  particulately fluidised bed, 14–15
  regular array of spheres, 14
drift, due to motion of sphere, 54
drops, 82–3
  rising velocity, 26–7
drying, 2, 122
dumping from sieve trays, 62
dye injection, 73–4
dynamical similarity, 120–1

eddy diffusion, see diffusion
effective mass, see inertia
empirical correlations, 120–1
emulsion, 113–114; see also particu-
    late phase
entrance effect on injected bubble, 31,
    45, 118
equilibrium with volatile particles,
    121–2

equivalent bubble diameter, see bubble
    diameter
equivalent flow through bubble, 101,
    106; see also exchange
eruption of bubbles, see bubble sur-
    facing
ethylene, 107, 113–14
excess flow as bubbles, see incipient
    fluidisation
exchange between bubble and particu-
    late phases, 20, 61, 63–79, 83, 106,
    113, 116, 119–21
  high rate, 103
  theory, 69, 94–106
  see also flow through void
expansion
  aggregative bed, 29, 99–100, 109,
    116
  liquid column, 29
  particulate bed, 15–19, 80
  see also bubble expansion, and height

fabric, drying of, 3
fast reaction, see conversion
film, high speed, see photography
filters, 65
fingers of particles, 79
first-order reaction, 97–118
  experimental results, 106–15
Fischer–Tropsch process, 2
fixed bed of particles
  compared with fluidised bed, 16–
    19
  drag force on sphere in, 14
  heat and mass transfer, 121
  oxidation of ammonia in, 114
  ozone conversion in, 107
  percolation, 71–2, 129–31
  pressure: distribution, 65; drop,
    9–12, 18
  voidage fraction, 10, 129
flow
  as bubbles, see incipient fluidisation
  through bubble, see exchange
  through void, 69, 95–6, 130–1;
    see also exchange
  within bubble, see circulation
fluorination of uranium, 2
formation, see bubble formation
form drag, 13
free-falling velocity of particles, 1, 16,
    84–5, 93
frequency of bubbles
  in a fluidised bed, 30, 48, 57–60,
    62
  in inviscid liquid, 50

friction coefficient
packed beds, 10–12
pipes, 9
friction factor, 15
frontal diameter, *see* bubble diameter
Froude number, 90–1
full-scale plant, 98, 121
full-size reactor, *see* large fluidised
beds

γ-ray method, 19, 30, 60
gas bubble, 21
diffusion within, 134–7
in a fluidised bed, 30, 34–7, 93;
forming, 57–62
in a liquid, 27, 31–2; forming, 50–
6
gas film resistance, *see* bubble, transfer
coefficient, diffusion, and mass
transfer
gas-fluidised bed, 1, 3–8, 34–8, 40–2,
48, 54, 61, 74, 80, 82, 84, 87, 89,
107
gauze, 1, 7, 62, 97
glass
beads, 30, 34, 75, 80, 110; bubble
velocity in, 31, 35; fluidised by:
air, 4, 30, 74; paraffin, 81, 89, 91;
water, 4, 18, 80
spheres, 41
glycerol, aqueous, fluidising lead shot,
81, 89, 91
grid, *see* porous plate, and gauze

heat transfer, 1, 8, 121–2
height of bed, 3–4, 29, 42, 44, 113
at incipient fluidisation, 3, 29
effect on: bubble diameter, 42–3,
48–9, 59, 99, 109–12, 114–16,
119–22; catalytic conversion, 103,
108, 111, 117
see also expansion
height of bubble, *see* bubble height
height of liquid, 29
helium, 107
injection, 115–17
Hill's spherical vortex, 136
Hirschfelder's method for estimating
D_G, 107
hollow resin, *see* resin
hydraulic mean diameter of passage,
9–10
hydrocarbons, 2
hydrogen, 107, 113
bubbles formed in water, 53
hydrogenation, *see* ethylene

hydrostatic pressure, effect on bubbles,
43
hysteresis in pressure drop curve, 3

ideal fluid, *see* inviscid liquid
incipient fluidisation, 50, 102, 115
bubbles account for excess flow, 29,
42, 61–2, 64, 85, 98, 100, 116
condition for, 67, 75
definition, 1, 3, 8–15
importance in two-phase theory, 19
viscosity of the bed, 40
incipient fluidising velocity, 8–15, 61,
67, 115
compared with bubble velocity, 41,
71–4, 94–6
measurement, 8
prediction, 13–15, 113, 118
incipiently fluidised bed, 73
density of, *see* density
formation of bubbles, 57–61
injection of carbon tetrachloride,
117–18
liquid-like behaviour, *see* analogy,
and two-phase
single bubble: velocity, 30–4, 40–1;
theory, 65, 74, 94–6
incomplete fluidisation, 111
incompressible fluid
fluidising fluid treated as, 65
particulate phase treated as, 64, 75,
79
theory, 123–9
inertia
air within bubble, 52
fluid round a bubble, 75
forces: bubble in a fluidised bed,
40; drops and bubbles, 21; fixed
beds, 12; particle motion, 75, 77
liquid round a bubble, 50, 52, 56,
128
particulate phase round a bubble,
60, 75
injected bubble, 30–1, 36, 43–4, 74–5,
82, 89, 92, 117–18
volume: compared with bed expansion, 36; lost on injection, 37
injection orifice, *see* orifice
instability of the lower surface of a
fluidised bed, 62, 81
interchange, *see* exchange
interfacial tension, 26
inter-particle
forces, 67, 79
pressure, 66–7, 74–9
interstitial velocity, 65–9, 75, 129

inviscid liquid
bubble formation in, 50–3, 56, 60
bubble rising in, 22–6, 32, 66
equations of motion, 123–8
particulate phase behaves as, 20, 50,
65, 79; *see also* analogy, and two-
phase
ion-exchange resin, *see* resin
iron-oxide particles, 99, 103, 107, 109
irregular bubble formation, *see* bubble
formation
irrotational motion
bubble, 24, 70, 72, 74
cylinder and sphere, 123–8
isolated sphere or particle in a uniform stream, 9, 12–17

jet of air or gas, 51, 55, 61

kinematic viscosity, fluidised bed,
40
kinetic energy due to a moving
sphere, 128

Laplace's equation, 126–7, 129
large fluidised beds, 8, 103, 108, 112,
115–17, 119–21
incipient fluidising velocity, 8
lead shot, fluidised by
air, 82, 89, 92
glycerol, 80–1, 91
paraffin, 81, 88–9, 91
water, 5, 37, 73–4, 80, 82, 89, 92
leakage
from bubble, *see* bubble formation,
and exchange
particles through a distributor, 62
level of liquid, *see* height of liquid
light probe, 30
light transmission, 19
limiting size of bubble, *see* bubble
diameter
limit of penetration, *see* penetration
liquid-fluidised beds, 1, 4–6, 8, 37, 80,
82, 84, 87, 89
liquid-like character of fluidised beds,
see analogy, inviscid, and two-
phase
long bubble, *see* bubble, rising, in a
tube
longitudinal mixing, *see* mixing

magnetite, 110
mass transfer, 79, 119, 121–2
coefficient, 112, 137
within bubble, 134–7

see also bubble, transfer coefficient,
and transfer coefficient
material balance, 123–4
bubble: rising, 100, 104; within,
134–5
infinitesimal bed height, 104
particulate phase, 64, 101–2
maximum size of bubble, *see* bubble
diameter
micro-balloon particles, 80, 89–90
minimum fluidisation velocity, 14; *see
also* incipient fluidising velocity
mixing, 117
bubble wake and particulate phase,
74
due to bubbles, 48
particulate phase, 98, 100–5, 109
streams leaving bubble and particulate phases, 102
see also complete mixing, and
piston flow
molecular diffusion, *see* diffusion
momentum
jet, importance in bubble formation,
50–1
liquid surrounding bubble, 52
multiple bubble formation, *see* bubble
formation

naphthalene, 121–2
Newtonian fluid, 40
nitrobenzene, 24
nitrogen dioxide, 31, 71, 74–5
nitrous oxide, 115
nose of bubble, 22, 24
number
bubbles per unit volume, 29, 100,
102
spheres or particles per unit packed
volume, 11, 15, 102

orifice, 48
bubble formation at: fluidised beds,
49–50, 57–61; two-phase systems,
50–6
oscilloscope, 57
overall conversion, *see* conversion
oxidation of ammonia, 114–15
oxygen, 107–8, 112, 114
ozone, 99, 103, 107–9, 122

packed bed, *see* fixed bed
paddle viscometer, 40
paraffin, 81, 88–91
particle diameter, 110
effect on: aggregative and particu-

late fluidisation, 90–1; bed viscosity, 40–1; bubble diameter 99; bubble stability, 85–9; bubble velocity, 31, 34–5; free-falling velocity, 85; mass transfer, 99, 112–13

particle motion, 30, 65–7, 70, 72, 74, 94; *see also* particle velocity

particle size distribution, 88–9

particle velocity, 8, 64, 68, 70, 74–5, 100; *see also* free-falling velocity

particles, 62
  agglomeration, 2, 7, 89
  catalytic, *see* catalytic
  fine, or small, 7, 72, 89, 92–4, 113
  fixed bed, *see* fixed bed
  fluidised, behave as liquid, *see* analogy
  free-falling velocity, *see* free-falling velocity
  incipiently fluidised, *see* incipiently
  large, 8, 59–60, 72
  terminal velocity, *see* free-falling velocity
  velocity, *see* particle velocity, *and* bubble velocity

particulate fluidisation, 6
  definition, 4–5
  expansion of particulate bed, *see* expansion
  relation to aggregative fluidisation, 4–5, 80–2, 85, 89–91, 94, 122

particulate phase, 19, 60–1, 63–79, 94, 96–9, 122
  concentration within, *see* concentration
  density of, *see* density
  diffusion within, *see* diffusion
  exchange of fluid with bubble phase, *see* exchange
  mixing within, *see* mixing
  motion of fluid and particles in, 63–6, 70, 72
  treated as an incompressible, inviscid fluid, *see* incompressible, inviscid, analogy, and two-phase
  velocity, 100
  velocity constant within, *see* velocity constant
  viscosity of, *see* viscosity

penetration
  circle of, 71
  limit of, 73–4, 96

percolation, 42, 65, 72, 77, 129–31
  through: cylindrical void, 71, 129–30; spherical void, 130–1

perfect mixing in particulate phase, 100–3; *see also* mixing

perforated plates, *see* sieve trays

permeability constant, 64

petroleum, 2

phenolic micro-balloons, *see* micro-balloons

photography, 30, 38–9, 49, 55, 64, 71, 73, 82–3, 92

pilot plant, 112, 119

pipe flow, 9

piston flow, 110
  in particulate phase, 98, 103–5, 113–14

plastics, 80

pneumatic transport, 6, 20, 37

Poiseuille equation, 11

porous plates, 7, 28

potential flow, 22, 24; *see also* velocity potential

prediction of incipient fluidising velocity, *see* incipient fluidising velocity

pressure
  equivalent to inter-particle forces, *see* inter-particle
  in: bubble, 21, 66, 76, 133; drop, 26; fluid flowing round spherical-cap bubble, 132–3; fluidised bed, 63, 66–7, 74–9, 113–14; fluidising fluid, 64–8, 74–9, 90; irrotational motion, 123, 126; percolating fluid, 129–31; wake, 23, 26, 132–3

pressure distribution, 24, 65
  around a rising bubble, 64–7, 69, 74–9, 132–3

pressure drop in fluidised beds, 3–4, 9, 12–13

pressure gradient
  at infinity, 67, 75–6, 129–30
  fixed bed, 11, 129–30
  in fluidising fluid, 64, 73

pressure recovery in wake, 132–3

propylene, 107, 111

pseudo-first order reaction, 114

pyrites, 2

quadruple bubble formation, 55

quality of fluidisation, *see* aggregative fluidisation, and particulate fluidisation

quartz, 40

quiescent bed, 30–1; *see also* incipiently fluidised bed

radiation, 30

radius
    bubble, *see* bubble radius
    circle of penetration, *see* penetration
range of fluidised state, 5–8
rate of exchange between bubble and
    particulate phases, *see* exchange
rate of rise of bubbles, *see* bubble
    velocity
reaction, *see* ammonia, ethylene, cum-
    ene, nitrous oxide, ozone
    catalytic, *see* catalytic
    velocity constant, *see* velocity con-
        stant
reactor, *see* catalytic
reduction of uranium, 2
regenerator, *see* catalyst
relative velocity between bubbles, *see*
    bubble velocity
resin particles, 5, 81, 88–9, 91, 110
Reynolds number
    bubble, 40–1
    fluidised bed, 16–18, 65
    packed bed, 9–12, 65
    sphere, 12–13
rise of bubbles, *see* bubble, rising
roasting pyrites, 2

sand, 3, 31, 34–5, 40, 43–4, 48, 65, 80
scaling-up, 98, 119–21
screen, *see* gauze
sedimentation, 16
seeds, swede, 31, 34–5
semi-angle, *see* bubble, spherical-cap
shallow beds, 121–2; *see also* height
shape of bubbles, 96; *see also* bubble,
    spherical-cap, angle
shear, 9, 82, 123, 125
short circuiting, 131; *see also* flow
    through void, and exchange
sieve tray, 51, 61–2
silica, 99
    gel, 122
single particle or sphere, *see* isolated
size of bubble, *see* bubble diameter
skin friction, 13
slug, 22, 27, 36
slug flow or slugging, 27, 32, 37, 41–2,
    109, 111, 114–15, 120–1
    definition, 5–6
    reduction, 7
small bubbles, *see* bubbles, stream
smooth fluidisation, 7, 80–1, 85,
    87–90, 93; *see also* particulate
    fluidisation
solids flow through perforations, *see*
    leakage

sphere
    force on, 12, 14
    irrotational motion past, 25, 123,
        128
    isolated, in uniform stream, *see*
        isolated sphere
spherical-cap bubble, *see* bubble,
    spherical-cap
spherical void
    flow of fluid in, 94–6
    percolation through, 69, 129–31
    see also exchange
spouted beds, 20, 37, 61
    definition of, 6–7
stabilising fluidised beds, 7–8
stability of bubbles, *see* bubble sta-
    bility, and bubble wake, stability
stability of fluidised beds, 81–4
stagnation point, 22, 133
steady rise of bubbles, *see* bubble
    velocity
steel particles, 80–1, 89, 91
Stokes law, 12, 14, 17
stream function, 68, 70, 94, 124–31,
    135
streamline, 23, 70, 72, 124, 127, 129,
    134–5
streamline flow, 13
streamlines in fluidised beds
    comparison with experiment, 70–4
    fluid, 68, 70–4
    particles, 66, 70–4, 83
superficial velocity
    at incipient fluidisation, 3, 13–15;
        see also incipient fluidising velo-
        city
    fluidising fluid, 3–4, 19, 95, 108–9,
        111–12, 114–16
    through packed bed, 11, 131
surface area of bubble, *see* bubble
    area
surface tension, 21, 48, 50, 54, 93
swede, *see* seeds
symmetry about an axis, *see* axi-
    symmetric

Teeter beds, 8
temperature
    fixed beds, 107
    fluidised beds, 1, 107, 110, 113–14,
        122
tension, *see* interfacial, and surface
terminal velocity, *see* free-falling
thickness of bubbles, 30; *see also*
    bubble height
through-flow for bubble, *see* exchange

time
  coalescence of bubbles, 44–5
  formation of bubbles, 52
  interval between bubble injections, 43–5, 117
  rise of bubbles, 31, 33, 36
tracer experiments, 31, 71, 73–5, 99, 115–16
transfer, *see* heat transfer, and mass transfer
transfer coefficient
  between bubble and particulate phases, 98, 110–12, 114
  within bubble, *see* bubble, transfer, and mass transfer
transfer units, 108
transient experiments, 115–16
transition between aggregative and particulate fluidisation, 5, 80–2, 86, 89–91, 94, 122
transport between bubble and particulate phases, 120; *see also* exchange, and transfer coefficient
tube
  bubble in, 21–3, 35
  diameter, 22, 32
  stream of bubbles in, 26–8
  see also diameter
turbulence, 23, 25, 28, 132
two-dimensional
  fluidised beds, 30–1, 39, 43–4, 54, 63–71, 74, 79
  motion, 123–6, 129–30
two-phase
  systems, 29, 38, 83, 93
  theory of fluidisation, 19–20, 110
  see also analogy, and incipient fluidisation; application to bubble coalescence, 47–8

UOP catalyst, 110
uranium, 2
uses of fluidised beds, 1–3

velocity
  between fluid and particles, 64, 69
  bubble, *see* bubble velocity
  continuously generated bubbles, *see* continuously
  drops, 26–7
  fluidising fluid, absolute, 95
  see also interstitial
  incipient fluidisation, *see* incipient

interstitial, *see* interstitial
irrotational motion, 123–8
particles, *see* particle velocity, and free-falling velocity
percolating fluid, 129–31
relative, between bubbles, *see* bubbles
superficial, *see* superficial
wake, *see* bubble wake
within bubble, 83–4, 92–3, 96, 134
velocity constant of reaction, 98, 102–3
  oxygen, 112
  ozone, 107
velocity potential, 25, 66, 70, 126–30
vertically unmixed emulsion, 113; *see also* piston flow
viscometer, 40
viscosity, 9, 23, 26, 50, 85, 90–1, 123, 125
viscosity of fluidised bed, 40–1, 75, 79, 81
viscosity of fluidising fluid, 81
  effect on bubble stability, 89
viscous forces, 12, 21, 40–1
void, 37, 69, 82, 129–31
  volume in packed bed, 10
  see also cylindrical void, and spherical void
voidage fraction, 16–18, 30, 80–1, 90–1, 94
  fixed or packed bed, 10, 129
  incipient fluidisation, 3, 13, 19, 65
  minimum, *see* voidage at incipient fluidisation
volatile particles, 121
volume of bubble, *see* bubble volume, and bubble diameter
vortex, 136

wake, *see* bubble wake
wall effect, 31–4, 36, 39, 46, 118, 120; *see also* diameter
water, 53, 74; *see also* bubble, water
water bell, 82, 93–4
water bubbles, *see* bubble
water fluidised bed, 4, 18, 37, 73–4, 80, 82, 85, 88–9, 92
waxes, 2, 7

X-ray, 30, 38–9, 79

Printed in the United States
By Bookmasters